Computers in microbiology

a practical approach

TITLES PUBLISHED IN THE PRACTICAL APPROACH SERIES

Series editors:
Dr D Rickwood
Department of Biology, University of Essex
Wivenhoe Park, Colchester, Essex CO4 3SQ, UK
Dr B D Hames
Department of Biochemistry, University of Leeds
Leeds LS2 9JT, UK

Affinity chromatography
Animal cell culture
Antibodies I & II
Biochemical toxicology
Biological membranes
Carbohydrate analysis
Cell growth and division
Centrifugation (2nd Edition)
Computers in microbiology
DNA cloning I, II & III
Drosophila
Electron microscopy in molecular biology
Gel electrophoresis of nucleic acids
Gel electrophoresis of proteins
Genome analysis
HPLC of small molecules
HPLC of macromolecules
Human cytogenetics
Human genetic diseases
Immobilised cells and enzymes
Iodinated density gradient media
Light microscopy in biology
Lymphocytes
Lymphokines and interferons
Mammalian development
Microcomputers in biology
Microcomputers in physiology
Mitochondria
Mutagenicity testing
Neurochemistry
Nucleic acid and protein sequence analysis
Nucleic acid hybridisation
Oligonucleotide synthesis
Photosynthesis: energy transduction
Plant cell culture
Plant molecular biology
Plasmids
Prostaglandins and related substances
Protein function
Protein sequencing
Protein structure
Spectrophotometry and spectrofluorimetry
Steroid hormones
Teratocarcinomas and embryonic stem cells
Transcription and translation
Virology
Yeast

Computers in microbiology

a practical approach

Edited by

T N Bryant

Medical Statistics & Computing,
University of Southampton,
Southampton General Hospital,
Southampton SO9 4XY, UK

J W T Wimpenny

Department of Pure and Applied Biology,
University of Wales,
College of Cardiff,
Cathays Park,
Cardiff CF2 1XH, UK

IRL Press
Eynsham
Oxford
England

First published 1989

British Library Cataloguing in Publication Data

Computers in microbiology.
1. Microbiology. Applications of computer systems
I. Bryant, T.N. II. Wimpenny, J.W.T. III. Series 576′.028′5

Library of Congress Cataloging in Publication Data

Computers in microbiology: a practical approach/edited by T.N. Bryant and J.W.T. Wimpenny.
p. cm.
Includes bibliographies and index.
1. Microbiology—Data processing. 2. Microbiology—computer simulation. I. Bryant, T. N. II. Wimpenny, J. W. T. (Julian, W. T.)
[DNLM: 1. Computers. 2. Microbiology—methods. QW 26.5 C738]
QR69.D35C66 1989
576′.028′5—dc19

ISBN 0 19 963014 3 (hardbound)
ISBN 0 19 963015 1 (softbound)
Previously announced as:
ISBN 1 85221 084 2 (hardbound)
ISBN 1 85221 086 9 (softbound)

Typeset and printed by Information Press Ltd, Oxford, England.

Preface

Computers are now a part of our lives, almost as familiar as the telephone and the microwave oven. Microbiologists are not divorced from this general trend and their attitudes to this technological revolution is quite varied. Some have never programmed a computer in their lives and view the keyboard with a certain suspicion even now. Others are converts, happily exploring the power and the puzzle that computers provide. Many have used computers for most of their professional lives.

This book was written by microbiologists who are experts in their fields for other microbiologists. Hopefully each of the groups described above will gain something from these articles.

The areas covered in the book discuss the application of computers to a number of fundamentally important areas in microbiology. These include clinical microbiology, data analysis, fermentation measurement and control, image analysis, modelling and simulation, taxonomy and systematics, and finally teaching. Each chapter illustrates how computers can be used to solve microbiological problems and is not concerned with computer programming *per se*.

T.N.Bryant
J.W.T.Wimpenny

Acknowledgement

Trademarks/Owners

IBM/IBM Inc.
Apple, Macintosh/Apple Computers Inc.
Turbo Pascal/Borland International Inc.
Microsoft/Microsoft Corp.

All other brand and product names are registered trademarks of their respective holders.

Contributors

R.M.Atlas
Department of Biology, University of Louisville, Louisville, KY 40292, USA

S.Bascomb
Department of Medical Microbiology, Wright Fleming Institute, St Mary's Hospital Medical School, London W2 1PG, UK

N.Bratchell
Institute of Food Research, Bristol Laboratory, Langford, Bristol BS18 7DY, UK

T.N.Bryant
Medical Statistics & Computing, University of Southampton, Southampton General Hospital, Southampton SO9 4XY, UK

K.A.V.Cartwright
Public Health Laboratory, Gloucester Royal Hospital, Great Western Road, Gloucester, GL1 3NN, UK

P.J.H.Jackman
National Collection of Yeast Cultures, Division of Genetics and Microbiology, AFRC Institute of Food Research, Norwich Laboratory, Colney Lane, Norwich, Norfolk NR4 7UA, UK

H.J.H.MacFie
Institute of Food Research, Bristol Laboratory, Langford, Bristol BS18 7DY, UK

P.G.G.Miller
Department of Microbiology, University of Liverpool, Liverpool L69 3BX, UK

P.Morgan
Shell Research Ltd, Sittingbourne Research Centre, Sittingbourne, Kent ME9 8AG, UK

A.G.O'Donnell
Department of Agricultural and Environmental Science, The University, Newcastle upon Tyne NE1 7RU, UK

J.I.Prosser
Department of Genetics and Microbiology, University of Aberdeen, Marischal College, Aberdeen AB9 1AS, UK

M.C.Whiteside
Department of Biological Sciences, University of Warwick, Coventry CV4 7AL, UK

J.W.T.Wimpenny
Department of Pure and Applied Biology, University of Wales, College of Cardiff, Cathays Park, Cardiff CF2 1XH, UK

Contributors

Contents

Abbreviations

ADC	analogue-to-digital converter
CAI	computer-assisted instruction
CCD	charge-coupled device
CDSC	Communicable Disease Surveillance Centre
c.f.u.	colony-forming units
CPU	central processing unit
DAC	digital-to-analogue converter
DAPI	4′,6-diamidino-2-phenylindole
DDC	direct digital control
DEFT	direct epifluorescent filter technique
DFA	discriminant function analysis
DHA	District Health Authority
DOT	dissolved oxygen tension
DTP	desk-top publishing
ELISA	enzyme-linked immunosorbent assay
EM	electron microscopy
GLC	Gas–liquid chromatography
GLIM	Generalized Linear Interactive Modelling
HIV	human immunodeficiency virus
HPLC	high-performance liquid chromatography
IA	image analysis
ID	identification
K	Kilobyte
LAN	local area network
LAT	lock-up table
MANOVA	multivariate analysis of variance
Mb	megabyte
MFLOPS	million floating point operations per second
MIC	minimum inhibitory concentration
MIPS	million instructions per second
MLP	Maximal Likelihood Program
MPU	microprocessor unit
NLQ	near letter quality
OTU	operational taxonomic unit
PAD	packet assembler disassembler
PCR	principal components regression
PDL	page description language
PHLS	Public Health Laboratory Service
PID	proportional integral derivative
PIO	parallel input/output
PLS	partial least squares
RAM	random access memory
R–K	Runge–Kutta
ROM	read only memory
SSC	supervisory setpoint control
WORM	write one read many

CHAPTER 1

Introduction

T.N.BRYANT and J.W.T.WIMPENNY

1. INTRODUCTION

Computers are at the centre of a modern technological and information revolution. Their speed, power, flexibility and sheer value for money are changing almost every aspect of our lives, hopefully for the better.

Microbiology, like almost any other field of endeavour, stands to gain from this revolution. Microbiologists, as do scientists in other fields, vary in their response to the phenomenon of computing. A few enthusiasts are familiar with the theory of computer operation. They understand the operating system and are able to design and implement programs of varying degrees of sophistication. Others are excited by the possibilities of applying these techniques to their work but lack both the experience and the confidence perhaps to get started. Another group eschew the field for one of a number of reasons. This book is aimed at all these classes. We hope that experienced computer users will gain at least some information from these pages whilst the reluctant and not so reluctant debutante will find that taking the plunge is really not so hard.

The application of computers in any field roughly divides into two parts. First, general computing techniques not confined to any one discipline, and second, specific applications used mainly or wholly by practitioners in the field. Since the first area (which includes, e.g. word processing), can actually take up most of the scientist's computing time we plan to discuss these topics briefly in this introductory chapter. Specifically microbiological subjects will be treated separately in the relevant chapters.

2. WHAT IS A COMPUTER?

Many readers already have a working knowledge of computers and so may skip this brief introduction. If fuller details are needed the bibliography at the end of this chapter should be consulted. Most people have some idea of what computers are and what they can do, thus a computer is, 'a machine which, under the control of a stored program, automatically accepts and processes data, and supplies the results of that processing' (British Computer Society, 1).

2.1 **A simple computer**

The diagram (*Figure 1*) is a simple representation of a computer showing the hardware or the physical components that are needed. The essential element of any computer is the central processing unit (CPU). The CPU carries out two basic functions, it stores information and it acts as a calculator. The CPU is a very simple brain analogue and in the same way that the brain can only function in the context of the rest of the body,

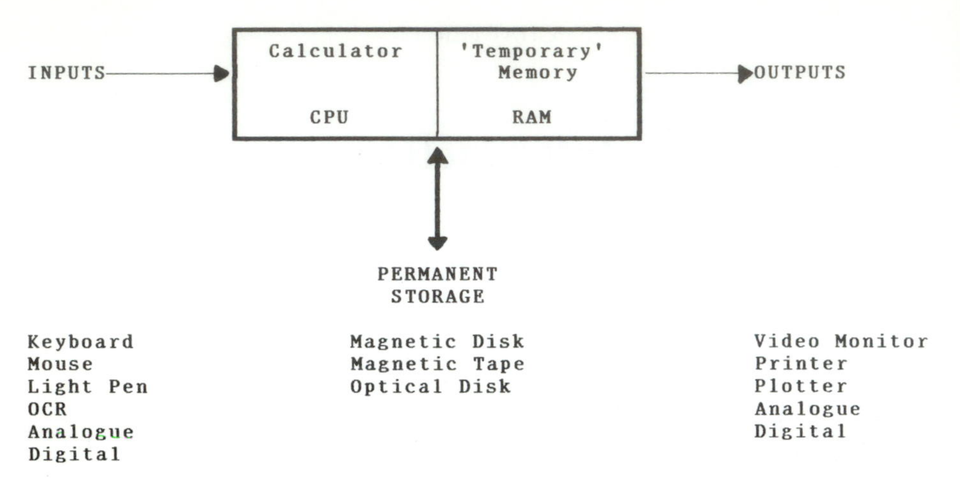

Figure 1. A diagrammatic representation of a computer.

the CPU requires 'peripherals' to perform useful work. The latter include the provision of power and the ability to communicate with a wide range of input and output devices. Although the CPU contains memory this is typically 'dynamic' and once the electrical power is switched off its contents are lost.

Permanent memory or 'backing store' is usually provided by magnetic disc or magnetic tape although optical discs based on laser technology are now available. Discs are either 'floppy' (also called discettes) or 'hard' (sometimes referred to as Winchester discs). Floppy discs are removable, lightweight, flexible structures that can be sent through the post. Hard discs are rigid and they operate within a sealed unit. With few exceptions they are not removable. Because they are rigid, hard discs have closer geometrical tolerances leading to higher packing densities and rates of data transfer that may be 10 times that possible with floppy discs. Optical media, though capable of storing gigantic amounts of information, are primarily for data that, once created, does not need altering. Such devices include WORM drives. The acronym stands for write once read many (times).

2.2 **Types of computer**

In principle the organization of the CPU is the same for all computers, be they micro-, mini- or mainframe. The classification of computer types is very blurred, the British Computer Society defines a mainframe computer as, 'a computer with a variety of peripheral devices, large amount of backing store and a fast processing unit. The term is generally used in comparison with a smaller or subordinate computer' (1). Clearly the terms used in this definition are relative since the modern microcomputer is as powerful as the mainframe system of 10 years ago. Perhaps a better definition of a mainframe computer is that it costs a great deal of money and it needs an air conditioned environment and operators to run it! The term minicomputers is vague and perhaps only persists now because historically some computer firms established a niche selling small laboratory-based computers which, to distinguish them from their costlier brethren, were christened minicomputers. Today both mainframe and minicomputers share another attribute. They are designed and optimized to share processor power between a number of different end-users or tasks. The power and the speed of these systems, allow each user his own 'virtual' computer even though many other users could be connected simultaneously. For some applications this concept of the virtual computer could be unsatisfactory. Thus a few processor-intensive numerical simulations or a system operating near its design capacity could become intolerably slow. Highly interactive applications such as word processing, especially using screen editing systems, become impossible to use properly under these conditions and perform much better on a microcomputer dedicated to a single user. As a result word processing programs on mainframe computers remain at the dinosaur stage in evolution! The mainframe computer comes into its own in a number of important areas. It can run lengthy numerical programs in batch mode as a background job or overnight; it offers message and file transfer and electronic mail facilities between its own users and between users at other sites throughout the world and last, but not least, it offers comprehensive archiving and backup facilities as a routine service.

One feature that distinguishes microcomputers from the rest is that they have become consumer durables traded like TV sets from high street stores. They have become desk-top artefacts not out of place in an office or a living room. They need neither sophisticated power supplies nor air conditioning and their computational power, however it is measured, is still increasing at an astonishing rate.

This rapid increase in power and availability means that a great deal of time and application is needed to assimilate the necessary information to use it to its full potential. On the other hand power and sophistication also means that 'user-friendliness' can be designed into operating systems and commercially available software so that comparative novices can achieve results quickly. This is quite in contrast to the situation on mainframe and on some minicomputers, where it is expected that computer staff will customize applications plus documentation for a generally computer literate population.

2.3 **Representing information in a computer**

The basic element used to hold information in a computer is the binary digit or bit. Like a switch this can have either of two states, off or on, represented in binary form as 0 or 1. Computers manipulate collections of bits. A group of 8 bits is known as a byte which represents 256 combinations ranging from 00000000 to 11111111 (0–255 in decimal). Since it is tedious to work in binary notation the hexadecimal notation using numbers to the base 16 is used instead. Here numbers up to nine are represented by 0–9 as usual whilst 10–15 use the letters A–F. This reduces the number of digits used to express a number and is more appropriate than decimal notation especially when dealing with bytes and binary data. Thus the decimal number 197, is 11000101 in binary and C5 is hexadecimal. Information may be manipulated as a single byte or as a combination of bytes called a word. The size of computer word can vary according to the type and make of mainframe systems where the word may be 32, 36 or even 60 bits long. Computers can obviously manipulate textual information, the 256 possible permutations of 8 bits are more than enough to allow a single byte to represent upper and lower case letters, digits, punctuation marks and so on. The most common code used to represent characters is called ASCII, or American Standard Code for Information Interchange. ASCII provides an easy way, although not the only

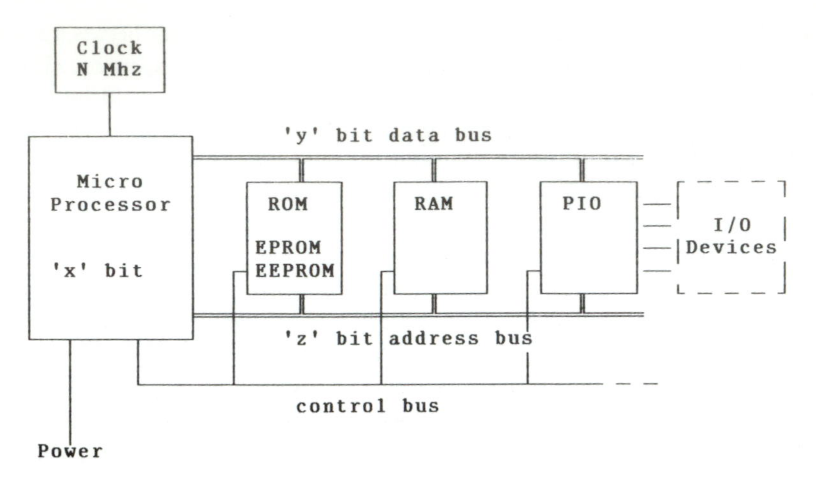

Figure 2. The main components of a microcomputer.

one, of allowing computers and peripheral devices to exchange information. Storage capacity in computing is frequently expressed as thousands of bytes or kilobytes, this is not exactly 1000 but 1024 bytes in decimal since it actually corresponds to the binary expression 2^{10}. Computer capacities are therefore larger than one might at first expect: for example 640 kilobytes (K) of RAM actually means 655 360 bytes of memory whilst 20 megabytes quoted for the capacity of a disc means that 21 309 440 bytes can be stored on it.

2.4 **A typical microcomputer**

The main components of a 'typical' microcomputer are shown in *Figure 2*. In principle the organization of larger computers is quite similar, though commonly several components are used to create the functional equivalent of a single component in a microcomputer. A typical microcomputer is based around a microprocessor unit, or MPU, which is a CPU without the memory, manufactured as a single integrated circuit. The activities of the MPU are synchronized by a clock whose frequencies range from 1 MHz upwards, depending on the type of microprocessor. All systems have a limited amount of permanent ROM or read only memory: this contains instructions to bring the machine to life when it is first switched on (booting it up in computer jargon). It usually provides some basic operating facilities such as the ability to delete or print information and a language such as BASIC in some systems. The information in ROM cannot be altered. A variation of this is the EPROM (erasable programmable read only memory) and sundry other similar devices which may be present in some systems. RAM, standing for random access memory, is the main memory in a computer. Its contents are volatile and lost when the power is switched off. Ram memory is used to store programs and all the necessary data required for the program to operate. In many computers it also contains the operating system (a specific set of programs) which is read into memory from a disc on booting up the machine. Input and output devices communicate with the MPU using a parallel input/output device or PIO and it is the latter which provides the link to the outside world. The transfer of information between these components is via a 'bus', or data highway. The bus has three functional parts, the data bus is bi-directional and transfers information between components, the address bus indicates where data should be taken from or transferred to, whilst the coordination of all these processes is the responsibility of the control bus. In the diagram the microprocessor, data and address buses are referred to as 'x, y and z' bit respectively to emphasize that the amounts of information that each bus uses may be different. For computer applications, microprocessors are available with 8-, 16- or 32-bit capacities: this allows up to 256, 65 536 and 4294 868 000 possibilities for the complete set of instructions for each size of microprocessor, respectively. The number of bits that the microprocessor can manipulate is not the only aspect to consider when viewing the capabilities of a microcomputer. To point to a single byte within the memory of the computer, the bit pattern for its location is placed on the address bus. This will point to a location which may be ROM, RAM or the PIO. The maximum amount of directly accessible memory is limitied by the numbers of bits used on the address bus. Most 8-bit computers use a 16-bit address bus giving access to 64 K of memory. The earliest microcomputers like the Apple II, the BBC microcomputer or PETs were based on 8-bit microprocessors: these had an 8-bit data and 16-bit address buses. The next generation of machines used 16-bit processors, however the width of the data and address buses varied from system to system. The IBM PC, based on the Intel 8088 MPU has an 8-bit data and 20-bit address bus whereas the Apple Macintosh based on the Motorola 68000 microprocessor has a 16-bit data bus and a 24-bit address bus. The latest systems such as the IBM PS/2 Model 80 or the Compaq 386 based on the Intel 80386 chip, and the Apple Macintosh SE, which uses the Motorola 68020 processor, are true 32-bit systems with 32-bit data and address buses. The evolving 'family' of computers is illustrated in *Figure 3*.

The clock is another vitally important component as events are synchronized by its pulses. In general, the faster the clock speed the faster the computer processes data; however the number of clock 'ticks' required for a complete operation to be performed is also important. If one microprocessor running at 2 MHz manipulates a piece of data in a single instruction whilst another running at 4 MHz needs two instructions to carry out the same manipulation then the net effect is that they both work at the same rate for this procedure.

3. CHOOSING A COMPUTER

Two things are worth remembering when choosing a computer. First, whatever is purchased will be 'old technology' within a short time. However, unless one is particularly image conscious, this may not be a problem as long as the system does what is necessary. The second point is not to underestimate requirements. Try to think what will be required of the system in 2–3 years time rather than solely what is needed now. Whilst this may constitute an activity akin to long range weather forecasting (and about as accurate!) it should in general mean buying more capacity at every level than one regards as strictly necessary. Early microcomputers had a maximum memory size of 64 K whilst the next generation typified by the IBM PC and its compatibles were given 640 K of directly accessible memory, however within 2 or 3 years, this limit, thought at the time to be reasonable, was too small for some applications. The systems

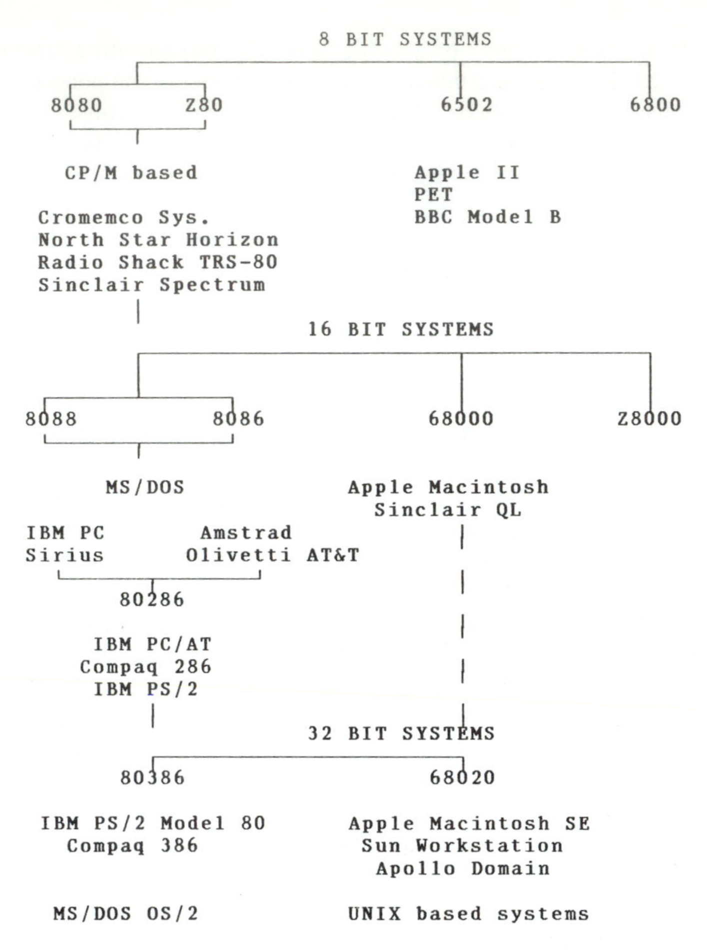

Figure 3. Relationship between families of microprocessors and microcomputers.

that replaced the IBM PC, the IBM PS/2 (Model 30-286 and above), were designed to permit users to access 16 megabytes of memory.

It is likely to cost more to upgrade a system at some stage in the future than to purchase a larger system in the first instance since the cost of an ‘add on’ is typically more than the cost of the component when supplied as part of a system. Having said that, it is also true that third party manufacturers often provide compatible components at lower cost, if one is prepared both to shop around and to accept the possible extra hassle involved in mixing suppliers. Though it may often seem intuitively silly, the major consideration in choosing a system is whether it will run the necessary software properly. To put this in a proper perspective it should be clearly understood that a computer with no software is an aesthetically pleasing lump which does little more than hum quietly to itself when switched on. The equation hardware + software = dedicated machine is true if not always obviously so. That the *potential* for the computer to become a word processor, fermenter controller, teacher, gel-scanner analyser, repository for bibliographic references, communicator for electronic mail etc. is there, is not in

Table 1. A basic desk-top computer system.

Memory	640 K RAM
Disc storage	20 Mbyte hard disk
	1 Mbyte floppy disc
Graphical display	640 by 350 picture elements (pixels), the colour version is preferable
Dot matrix printer	150 characters per second print speed, near letter quality (NLQ) preferred
Processor speed	At least twice the basic IBM PC speed
Optional extras	
Mouse	Graph plotter (six colours)
Tape backup machine	Laser printer
Maths coprocessor	

question. All it needs is appropriate software. That the second half of the dedicated machine is represented by a combination of 0s and 1s on a floppy disc may seem surprising, especially when the cost of these magnetic imprints can often be similar to the cost of the hardware. For instance the software needed to convert an Intel 80386-based machine to perform image analysis can be from 50 to 100% of the cost of the computer . . . .! The message is clear: select the necessary software first, then buy the computer that is best able to run it.

3.1 Suggested minimum configuration for a complete microcomputer system

Given that the necessary software will run on it, it is clear that most people will choose a microcomputer system that is compatible with the *de facto* standards imposed on the industry by IBM. This is not to say that the IBM PC, its numerous clones and its successor the IBM PS/2 are better. In fact rather the opposite may be true, thus many people regard the microprocessor and the design of the Apple Macintosh family as superior to IBM equipment and clones. On the other hand the IBM family have access to by far the largest range of software and must be regarded as 'safe' rather than an imaginative investment. *Table 1* gives one suggestion for a basic desk-top computer system.

As the capacities of hard discs increase and their prices fall, it becomes more and more important to copy precious material onto other media in case of accidents to the hard disc. 'Backing up' data is a matter of routine protocol by the system operators of mainframe and minicomputers. It *ought* to be just as important to the owner of a microcomputer. The fact that this is often not the case is due in some cases to inexperience and in others to laziness among the more experienced owners of most microcomputers. In the basic system described in *Table 1*, a large capacity floppy disc drive is recommended because it then becomes relatively easy to make backup copies of data on the hard disc. This will be simpler if work is divided into sections on the disc called sub-directories. Each sub-directory can then be copied onto one or a few discs. A tape streamer is the alternative method. Here the contents of the entire hard disc are copied at regular intervals onto magnetic recording tape. Tape streamers can be purchased which are capable of copying all the data from the hard disc onto a single tape. They naturally add to the cost of the system, however for serious work they must represent an excellent if unexciting investment.

The range of monitors available for use as computer displays is now very wide and the only problem is whether to choose colour or monochrome. Whilst a colour display is not essential it is generally preferable. The exceptions are where very high resolution is needed, for example in desk-top publishing, for complex graphical plotting and also in some areas of image analysis. Choice of monitor is really an item which is dictated both by application and by budget. On the whole colour monitors are three to five times dearer than the basic monochrome alternative. Whilst manufacturers continue to improve the resolution of colour monitors, reasonable values are 640 × 350 pixels. The most versatile (and expensive) monitors are so called 'multi-sync' devices which can operate with all the main graphic standards currently in use, monochrome, CGA, EGA, VGA etc. Because of this they are often advertized as being 'future proof' . . . . .!

Numerous types of printer are available for producing what is known in the trade as 'hard copy'! The most versatile choice is the dot matrix printer, though laser printers at about four times the price are becoming an economic alternative. Most dot matrix printers and all the laser printers are able to display graphical images and to produce near letter quality (NLQ) printed output. Dot matrix printers originally produced an image using 9-pin heads. The better quality modern printers now use 24 pins for increased clarity and resolution. Printers are usually equipped with some memory to store at least part of the material to be printed. If this 'printer buffer' is large enough to hold most or all of the information to be printed it can speed up operations allowing the user to get on with other work whilst the printer is in action. Seperate printer buffers containing more memory may be purchased. These sit between the computer and the printer and represent a useful 'extra' especially if the printer is used often or the device allows sharing between a number of computers. If the computer has sufficient spare RAM some of this can also be configured as printer buffer.

A 'mouse' is a hand held device whose movement over a hard surface is followed by a moving pointer on the computer screen. The mouse is equipped with one or more buttons which when 'clicked' activate a program or part of a program to which the arrow on the screen points. The mouse naturally cannot be used for entering data (typing a document etc.) however it does make the system more 'user friendly'. An experienced user can often work faster without a mouse, but for the inexperienced or casual user these creatures may be worth considering.

For application that are numerically intensive, for example mathematical modelling or image analysis, it may be useful to include a maths co-processor to improve program speed. This chip performs mathematical calculations in hardware as opposed to using software routines which are inherently slower. Some software may actually require such a co-processor whilst other programs can detect its presence or absence and act accordingly.

The days of the graph plotter may be numbered since high quality laser printers are capable of producing publication quality output better than all bar the most expensive plotter. The main advantages of graph plotters are their ability to use different coloured inks and their value in creating overhead transparencies.

3.2 **Sources of supply**

No one can resist a bargain, however, purchasing a computer or software from a supplier who advertises the lowest price can sometimes be a false economy. This is because such a supplier may be simply in the business of turning over items as fast as possible without either the desire or the ability to offer appropriate backup. The latter includes (where necessary) installation, instruction, maintenance and emergency service. In some cases appropriate support is available from one's own institution, and their advice should be sought at an early stage. They may even have special purchasing agreements with some computer manufacturers and generous discounts, perhaps as much as 40%, are often available. One or two firms are offering very good deals by selling directly to the end-user and at the same time offering very good technical backup and warranty terms.

4. SOFTWARE FOR MICROBIOLOGISTS

Where does one find appropriate computer software? This, of course depends entirely on the application in mind. Software for general non-microbiological applications such as word processing, databases, spreadsheets and the like are all extensively advertised and reviewed in the general computing literature. Often local computer centres will have a view as to the best package to buy. As stressed earlier, they may even have deals going with the supplier to obtain worthwhile discounts on certain items.

4.1 **Specific software for microbiologists**

There is rather little commercially available software for microbiologists. It is not hard to discover the reason for this. Software is extremely expensive to write professionally. It is only where there is a large market that it becomes economically worthwhile to write programs that can be sold at a price that makes money for the writer yet is cheap enough to be purchased by the user. The main exception perhaps is software for genetic engineering applications. This has become a popular high cost/high value area and users can generally afford expensive software since the latter plays a vital part in modern molecular biology.

There are four main approaches to obtaining relevant software.

(i) Professionally presented commercial programs.
(ii) Amateur programs published in the literature or available as supported from user groups or over electronic networks.
(iii) Unpublished programs obtained from a third party or unsupported programs available from user groups and over electronic networks.
(iv) Do-it-yourself programs.

Buying a program may seem to be the most expensive option, however this is too simplistic an answer. Writing a program can take many hours and only the individual can decide what this time is worth. Whilst there may be satisfaction in facing the academic challenge of developing software, there are two good reasons to question the value of doing this besides simply the hours of work involved. First, is one simply re-inventing the wheel? Second, can one expect, *however long it takes*, to produce programs that offer all the features of a commercial program or package? Naturally this depends on the sophistication of the package. An image analysis (IA) package (mentioned earlier) is an extremely sophisticated offering and it would be stupid to attempt to emulate this on one's own. On the other hand simpler programs may be worth the effort.

Table 2. Some commercial software of particular interest to microbiologists.

Title	*Computer*	*Producer*	*Cost*	*Comment*
Bacterial Identification	BBC IBM PC	IRL Press[a]	£65 $130	Teaching program for identifying bacteria.
Microcomputers in Biology	BBC IBM PC	IRL Press	£50 $100	Over 50 BASIC programs covering techniques used in Biology. Part of IRL Press book of same name[b].
Molecular Genetics	Apple Mac	IRL Press	£75 $135	Simulations of gene cloning experiments.
MicroModeller	BBC IBM PC	IRL Press	£65 $130	General purpose model building program. Good for biologists.
Enzpack	BBC IBM PC Apple II	Biosoft[c]	£40 $80	Single substrate enzyme kinetics.
Assay-Zap	Apple Mac	Biosoft	£125 $249	Universal assay calculator/curve fitting program.
Enzfitter	IBM PC	Biosoft	£80 $140	Enzyme kinetics package.
Molgraf	IBM PC Apple II	Biosoft	£80 $140	Draws molecules in 3-D.
PC-Taxon	IBM PC	COMPress[d]	$75	A taxonomic database.
Numerical Recipes	IBM PC	Cambridge[e] University Press	£22	Price is for discs in FORTRAN or Pascal of programs associated with a book of numerical analysis of the same name.
Yeast Identification Program	IBM PC	Cambridge University Press	£85	Takes user's test results and compares with ~500 known species in database.
Stella	Apple Mac	High Performance Systems[f]	$150 £199–£299	Visually orientated numerical modelling program. Excellent for biologists.

[a]IRL Press, Pinkhill House, Southfield Road, Eynsham, Oxford OX8 1JJ, UK
[b]Microcomputers in Biology: A Practical Approach (1985) Ireland,C.R. and Long,S.P. (eds), IRL Press, Oxford.
[c]Biosoft, 22 Hills Road, Cambridge, UK
[d]COMPress, PO Box 102, Wentworth, NH, USA
[e]Cambridge University Press, Cambridge, UK
[f]High Performance Systems, Lyme Road, Hanover, NH, USA

4.1.1 *Commercial programs*

Commercial packages in the area of microbiology are, as has been stressed, not very common. Such software as exists is advertized in the literature by firms such as IRL software, Elsevier Biosoft etc. A fuller list is included in *Table* 2. Much of this material is designed for teaching aspects of microbiology and at least some of it was in use some time before publishers became interested in marketing it commercially!

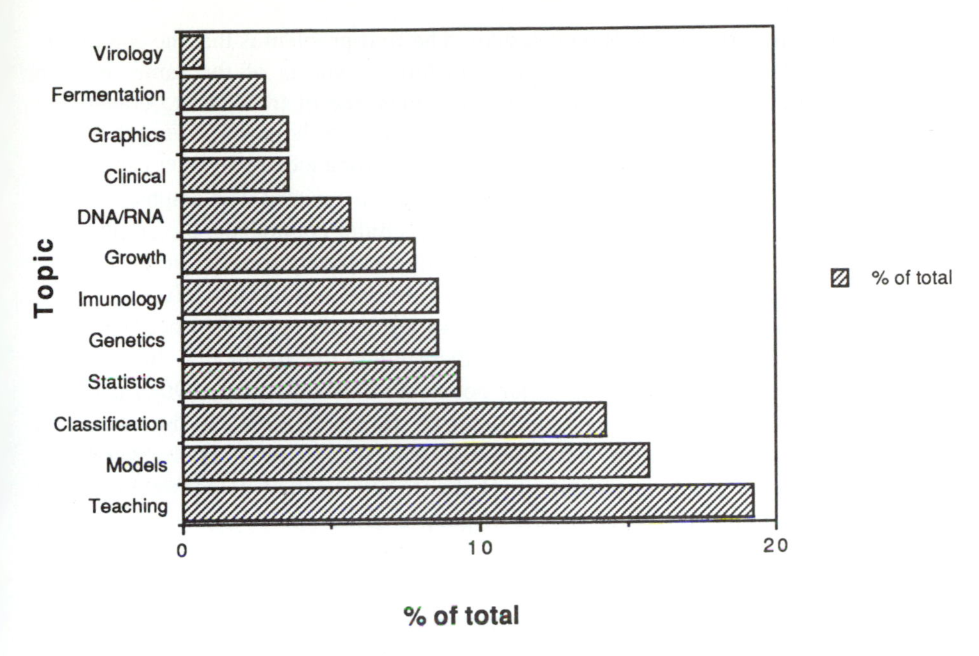

Figure 4. Topics covered in SGM Computer Club Program Catalogue.

4.1.2 *Amateur programs*

Published software or that available via electronic bulletin boards etc. may well serve its purpose admirably. The first problem is to track it down. Published software can be identified via a comprehensive literature search, although this may not always reveal what is there. This is because many programs are not considered to be a major part of the subject of the paper and as such do not feature either in the key words or the abstract of the paper.

Some help can be gained from indexes to journals which specifically relate biology to computing. A few organizations are establishing software catalogues. Thus the Society for General Microbiology (SGM) Computer Club publishes its Program Catalogue from time to time. This lists some 130 programs with details of their function and machines on which they run. Such programs can show a wide variation in quality. The main coverage of programs described in the Computer Club Program Catalogue are indicated in *Figure 4.* Some commercial programs are listed, however, the vast majority are written by microbiologists. They vary considerably in sophistication, in availability, in documentation and in program maintenance. If any of the descriptions seem interesting users are advised to get in touch with the author of the program directly. The Microbiology Computer Users Group (MCUG) of the American Society for Microbiology (ASM) also produce program catalogues and even maintain a program library although the latter consists mostly of general public domain software rather than specifically microbiological sofware.

4.1.3 *Unpublished programs*

It is possible to obtain software from third parties 'via the grapevine'. Whilst some of these may be ideal and work perfectly, it is more common to find that they will not, at least at first, run on one's own system. The first problem is that the computers are different. The program may use quite different dialects of the same or even completely different languages. The problem is then one of translation. Translating and de-bugging one's own programs can be tedious enough, however, trying to understand another writer's program well enough to rewrite it for another computer can be horrendously difficult. Particularly intractable are problems associated with graphics. Whilst a language such as BASIC may have many largely similar commands, the graphic and input/output capabilities of the machine are generally completely specific for one system. Finally, the program may not have been written properly in the first place . . . .! It is also well known that the authors of such software often do not either document their programs or even completely finish writing them. Only the author may be aware of the small fix he has to put into the program before it will run properly. Other people's software, often ill documented, can sometimes be a waste of time and the onus is on the user to check it carefully to decide whether it is suitable as it stands, whether it can be made suitable simply or whether it is better to rewrite the whole program, based perhaps on algorithms developed by the original author.

4.1.4 *Do-it-yourself programs*

Finally, writing programs: in spite of the caveats stressed earlier there is excitement and challenge in writing a program. Little advice can be presented here. The more competent reader knows the pitfalls and can make some attempt at predicting how much time it will take. To those with a growing knowledge of one or more programming languages one can only make the following observations.

(i) Choose a structured, portable language like Pascal, most versions of which allow programs to be transferred easily from machine to machine.

(ii) Plan the program in flow sheet form before writing a single line. Work out the algorithms involved right at the start. Plan, plan and plan again! This means staying away from the computer for as long as possible before starting to write the code. That the program writer will almost completely ignore this advice is sad but true!

(iii) Annotate the program as comprehensively as possible so that in the event of it being made available to another user it will be relatively comprehensible. Some programmers rule of thumb is one line of comment for one line of code!

(iv) When the program is nearly finished spare a thought for some kind of documentation that will allow others to use it easily.

(v) The program will take weeks to write, translate that directly into months—and this will still underestimate the investment in time.

(vi) When (if?) the program is finished add up the number of hours taken and multiply by some reasonable hourly rate of pay. Add in overheads, depreciation of equipment etc. to get some idea of the cost of your masterpiece!

(vii) If any of the following items apply, seriously consider the wisdom of embarking on writing programs: membership of more than two committees, the possession of a wife/husband, the possession of any children especially under the age of 10, a keen interest in sport, gardening, golf, highland dancing. . . . .! In other words almost all attributes of normality strongly mitigate against program writing!

4.2 **General purpose software**

4.2.1 *Word processing*

Perhaps the single most popular application of computers to professional life has been word processing. It is probable that microbiologists spend more time using word processors than they do in any other computer application, so that a brief description of the subject is worth including here.

The main difference between a word processor and a typewriter is that the latter is completely unintelligent. Where its 'moving finger writes and then moves on' in a singularly unreversible fashion there is a great deal of 'piety and wit' built into a word processor to undo or modify that which has been written!

The ability to modify textual data more or less parallels the history of computing. The line editors available (still!) on most mainframe computers as well as some of the more horrible editors like ED in CP/M or EDLIN in MSDOS are one end of a vast range of word-manipulating programs whose pinnacle might be the latest desk-top publishing (DTP) programs which bring the sophistication and professionalism of the printing trade to the keyboards of almost anyone interested enough.

Word processing consists of two fundamental activities. These are first text entry and its modification and second, the formatting of text in a form suitable for printing.

Text entry is almost always from a keyboard. The crudest systems as already mentioned, are 'line editors'. These allow the operator to move backwards or forwards through a text file on a line by line basis. Insertion of text and its subsequence correction can only operate on the current line. More modern mainframe computers and most microcomputers use 'screen editing'. Here the whole screen is 'active'. The cursor can be moved to any point on the screen using cursor control keys and corrections made at will. Editing can consist of the insertion or removal of text at any point in the file. This applies also to blocks of text which can be transferred individually to and from other text files, or copied from one part of a file to another by simple block operations. Two operations that are almost the essence of word processing are 'word wrap' and 'justification'. These are formatting operations which are used to produce the final printed text. A paragraph consists of a stream of text which flows into each line of whatever length. As a line fills up the word that would make the current line too long is carried over to the start of the next line, and so on. Each line can be justified, that is the right hand margin can be straight rather than ragged. To achieve this spaces are inserted between neighbouring words to pad out the text. Formatting commands are needed to tell the printer exactly how the page should be displayed. Such commands include specifying page length, line length and spacing, underlining, pagination, typing of repeated heading or footing text and so on.

There are numerous word processors from which to choose, depending on the microcomputer available. Even the simplest can be quite powerful enough to do real work. Two of the most scientifically productive colleagues of one author publish papers written exclusively on the 32 K RAM BBC micro using Wordwise. Wordstar (often described in the literature as the 'industry standard') was originally written for 8-bit machines running the CP/M operating system. It is now available for the IBM PC and all its clones. Wordstar is almost a WYSIWYG (what you see is what you get) program. This means that the appearance of the text on the screen is approximately as it will appear on the printed page. Missing on the screen are the results of printer control commands such as underlining, super- and subscript, doublestriking and so on. Other word processors are true WYSIWYG. Here text is often presented as black on a paper white screen. Moreover it also appears in one of a number of font styles and sizes and the system is generally manipulated using a mouse pointer device. Using the mouse block transfer (termed 'cutting and pasting') is very simple. This more sophisticated approach seems likely to be the 'standard' for the foreseeable future.

4.2.2 *Desk-top publishing (DTP)*

DTP systems overlap with, and take over from, word processing systems. Where word processors have their roots firmly in the realm of the typewriter, DTP systems are more akin to the printing press. DTP systems enable non-specialists to produce what are to all intents and purposes highly polished, professional publications. The latter can include almost everything from circulars, minutes, lecture notes, newsletters, magazines, newspapers and even books. . . . .!

The DTP process can be divided into a number of different activities. First input of information. DTP packages take textual data from the keyboard and from data files compiled by other word processors. They can also use graphical data from spreadsheets or graph drawing packages, collected and formatted information from database management systems and art work from drawing/painting programs.

The second activity is the design and implementation of the page layout. Many DTP systems allow one to store a customized layout so that it may be used again and again. The layout process is very similar to the art of the printer. It involves decisions on font type, style and point size for each level of heading and for the body of the text. Gaps between lines (leading) can be chosen as can the number of columns per page, framing around items and so on. Graphic items can be placed anywhere on the screen and the text allowed to 'flow' around it. DTP has become the most recent 'craze' on the microcomputer scene and many layouts can only be described as utterly tasteless for although the tools themselves may be excellent they, on their own, cannot make up for the art of the printer which has really to be learnt!

The final process is producing the printed page. There are a number of possibilities. The most common approach is to use a laser printer which can reproduce the screen image exactly. Laser printers have reasonably high resolution, around 300 dots per inch (d.p.i.). This is not quite as good as the resolution of the human eye (~ 1000 d.p.i.) but is suitable for most routine purposes and certainly for proofing items of work. The best solution is to have the final version printed out on an electronic typesetter. Thus the Linotron typesetting machines are capable of from 1200 to 2400 d.p.i. and the quality of the output is therefore impeccable. To communicate with typesetting machines a set of instructions are needed. These take the form of a page description language (PDL). There are several PDLs available. Perhaps the most useful is Postscript which is available for most commonly used microcomputers. For example output from the Apple Macintosh can be directed either to laser printer which is programmed in Postscript or to the Linotron typesetter machine which also understands Postcript. There are numerous introductory texts to DTP, for example Lang (2), Bate (3) and McLelland and Danuloff (4). The last mentioned is more of a catalogue of type and graphics and should be most useful to people planning any serious DTP work.

4.2.3 *Databases*

The term databases has different meanings to different people. One definition is, 'a collection of interrelated data stored together with controlled redundancy to serve one or more applications in an optimal fashion; the data stored so that they are independent of programs which use the data; a common and controlled approach is used in adding new data and modifying and retrieving existing data within the data base' (Martin, 5).

However a better working definition is, 'a database is a large organized store of information which is available when you need it' (Sharman, 6).

Computer programs used to create and permit the use of data stored in a database are referred to as *database management software*. The size and scope of a database can range from something analogous to a personal notebook, which might contain a list of names and addresses, to all of the clinical and administrative information commonly used by a large hospital, for example. The software packages available to manage such information vary in sophistication in a similar fashion.

The working definition of database does not actually mention the word 'computer' and in some cases such machines are not really needed. Thus advertisements for home computers suggest that one might store recipes in a database, however the latter are really far more trouble than a well written cook book.

Databases can be useful in many areas of microbiology. One example, stock culture collection information, illustrates this well. All the information about each culture could be written on an index card and the latter might be stored alphabetically according to the name of the organism. Should one also need the list stored by accession number then a set of duplicate cards should be made. If any details about a culture changed, then both sets of cards would need updating. Searching for organisms with particular attributes means examining each index card in turn, obviously laborious for a large collection. The production of a printed catalogue for the collection would entail transferring the information from the index cards, perhaps with some modifications, onto the printed page. This process in the absence of computers, was a tedious, error-prone and time-consuming chore, however with suitable database management software all of these things can be done very easily.

Broadly speaking there are two types of package that could be described as database management software. The simplest are really list processing programs. These are exactly analogous to the card index box, and like searching through one box generally operate on a single datafile. The usual approach to searching for data in these unsophisticated packages is to search through the entire data base each time a piece of information is needed. A refinement in some cases is the ability to support multiple indexes so that one can access a culture, as in our example above, either by name, or by accession number or, if such an index is set up, by source and site of isolation. There will inevitably be redundant information, thus all isolates of the same species should share the same optimum growth conditions (see *Table 3*). True database management software seeks to reduce redundant information to a minimum: to do this several files each holding part of the information describing a culture are generally employed. For further information on databases for microcomputers see Sharman (6).

Of special interest to academics perhaps is the bibliographic database manager which is a database system which has been optimized for storing literature references. There

Table 3. Storage of culture collection information.

1. Structure of database using a card index approach

Accession number	*Name*	*Growth temperature (°C)*	*Growth medium*
2046	*Aeromonas caviae*	25	Nutrient agar
2170	*A.caviae*	25	Nutrient agar
1366	*Alcaligenes faecalis*	37	Nutrient agar
2062	*A.faecalis*	37	Nutrient agar
669	*A.faecalis*	37	Nutrient agar

2. Structure of a relational database system: data is divided into two files or tables. The data in the table above can be created by linking the first table below with the second using the Name field.

Accession number	*Name*
2046	*A.caviae*
2170	*A.caviae*
1366	*A.faecalis*
2026	*A.faecalis*
669	*A.faecalis*

Name	*Growth temperature (°C)*	*Growth medium*
A.caviae	25	Nutrient agar
A.faecalis	37	Nutrient agar

is a wide choice of such software on the market at present. Some of these were reviewed (see Binary **8**, June, 1986 and *Table 4*). The better packages are extremely powerful although they may seem to be expensive when compared with the cost of popular word processing packages. Sadly the less powerful (and cheaper!) packages may not provide all of the most valuable facilities. One of the most useful attributes of a good bibliographic database is its ability to format references for printing out in the exact style of a particular journal or book publisher. This facility, combined with good searching routines embedded in a database which can manage a large collection of complete references, ought to be regarded as essential features to be really useful. Never forget that all the data has to be entered into the bibliographic database by someone. . . . .! This is not a step to be undertaken lightly, especially if one is committed to putting a large card index system onto the computer. It is perhaps more sensible to switch to a computer-based system for all new references plus whatever one can manage of the most recent index cards, than to attempt the more ambitious task. Attention should be paid at the outset to the design of the reference database. A mistake made at this stage could be disastrous.

An example of the sort of help given by bibliographic databases in manuscript preparation is shown in *Figure 5*. The first section of text shows the manuscript as prepared using a word processor, each reference uses an accession number assigned to it by a bibliographic reference manager. A second file is then created from the word processor file by a utility which substitutes the appropriate citation into the body of

Table 4. Some useful bibliographic database programs.

Title	*Computer*	*Producer*	*Cost*	*Comments*
REF-ED	IBM PC plus other	Skandigen Reference Systems[a]	£200	Up to 5000 refs, 512 characters in each.
Bibliographer	IBM PC	IRL Press	£49.50 $99	Up to 65 000 refs, 12 fields of 2 lines each.
Martz-Bibliofile	IBM PC plus CP/M plus others	Martz Software[c]	$250	Up to 32 000 refs, each ref. to 2000 characters.
Refsys Plus	IBM PC	Biosoft[d]	£99 $199	
Archivist	IBM PC	IRL Press[b]	£47.50 $95	Unlimited refs, 16 000 characters in each. Replaces Bibliographer.
Scimate	IBM PC	Institute for Scientific Information[e]	$399	From the people who publish Current Contents.
Xitat	Z-80 based	Proteus[f]	DM6000	50 000 refs.
Reference Manager	IBM PC Apple Mac	Research Information Systems[g]	$195 $440	RM-800 RM-32000

[a]Skandigen Reference Systems, Sweden; Infocard Nordiska AB, Sarogatan 2, S-16340 Spanga, Sweden
[b]IRL Press, Pinkhill House, Southfield Road, Eynsham, Oxford OX8 1JJ, UK
[c]Martz Software Inc., 48 Hunter's Hill Circle, Amherst, MA, USA
[d]Biosoft, 68 Hills Road, Cambridge, UK
[e]Institute for Scientific Information, 3501 Market Street, Philadelphia, USA
[f]Proteus, Haid-und-Neustrasse, 7–9, D-7500 Karlsruhe, FRG
[g]Research Information Systems, 1991 Village Parkway, Encinitas, CA, USA

the text and then creates a list of references used. The result of this process is shown in *Figure 5b*. If a different journal format is required this can be done literally at the press of a button as in *Figure 5c*.

4.2.4 *Spreadsheets*

Spreadsheets are electronic worksheets. They tabulate data within individual cells in columns and rows and can apply numerical relationships to specific cells or sets of cells. The rapidity with which they can calculate and re-calculate a matrix of data allows one to model numerical systems and to carry out 'what if' type experiments.

The history of spreadsheet programs and of the microcomputer are closely intertwined. From about 1979 onwards many commercial organizations purchased the Apple II microcomputer simply to run the program VisiCalc. When IBM introduced the PC in 1981, a new spreadsheet package, Lotus 1-2-3, was designed for it. The introduction of Lotus 1-2-3 ensured the success of the IBM PC and established microcomputers in the business community where spreadsheets are used for financial modelling. The scientific community has yet to adopt spreadsheets with the same enthusiasm.

However, they can be a useful tool and applications such as data acquisition and even

a The authors of this chapter have published in separate fields, One of us is interested in microbial responses to physical and chemical gradients (11) and the other in numercial taxonomic problems (127).

....

b The authors of this chapter have published in separate fields, One of us is interested in microbial responses to physical and chemical gradients (Wimpenny, 1981) and the other in numercial taxonomic problems (Bryant, 1987).

Bryant, T.N. (1987) Programs for evaluating and characterising bacterial taxonomic data. Computer Applications in the Biosciences, **3**, 45-48.

Wimpenny, J.W.T (1981) Spatial order in microbial systems. Biological Reviews, **56**, 295-342.

....

c The authors of this chapter have published in separate fields, One of us is interested in microbial responses to physical and chemical gradients (1) and the other in numercial taxonomic problems (2).

1. Wimpenny, J.W.T (1981) Biological Reviews, **56**, 295.

2. Bryant, T.N. (1987) Computer Applications in the Biosciences, **3**, 45.

Figure 5. Example of a bibliographic database, based on Reference Manager (Reference Information Systems Inc.). (**a**) Text as created using WordStar; 11 and 127 are the accession numbers of references in the database. (**b**) Text with references added and the references list produced in *CABIOS* format. (**c**) Format as used by IRL Press in this publication.

numerical modelling are being developed around them. In general, spreadsheets are based on microcomputers although some software is now available for larger systems.

The capacities of spreadsheet packages vary enormously; for example one might permit the use of up to 999 rows and 9999 columns. Others have a larger capacities and are well suited to long-term data capture and manipulation from on-line experiments.

The cell is a central concept in spreadsheet operation. Cells appear at the point of intersection of row with column identifiers. Three types of information may be appear in a cell.

(i) Textual information which is not further processed.
(ii) Numerical data entered by the operator.
(iii) Formulae for calculating cell contents using values from other cells.

A spreadsheet program can include many useful functions. Thus it is possible to count, and to calculate the total, mean and standard deviation of a set of values all by a single instruction. Many packages incorporate facilities for summarizing the data in graphical form. An elementary introduction to spreadsheets in a biological setting is described by Kibby (7,8).

Manufacturers of laboratory equipment together with scientific software publishers are now marketing data logging packages which can be linked to the most popular spreadsheets. One package called MAXIMA (Dynamic Solutions Corporation) which was designed for a range of chromatography instrumentation collects data which can be processed using Lotus 1-2-3. Similarly, the Beckman Liquid Scintillation Data Capture Software allows scintillation counter data to be transferred to spreadsheets or manipulated by programs written in a range of high level languages. Kellis (9) has produced a Lotus 1-2-3 template for enzyme assays using liquid scintillation data and the same company markets a data capture package to operate with their UV/visible spectrophotometers. These and other systems allow data files to be imported into a spreadsheet where it can be reduced, transformed, analysed and graphically summarized without entering a single element of data from the keyboard.

4.2.5 *Graphics*

Modern terminals, printers and especially microcomputers transfer textual information to the user in a visual fashion: it is not surprising then, that they often present graphical information in the same way.

Graphical software has been available for large computers for many years, however the exploitation of graphics by casual users required the development of the microcomputer. Large or mainframe computers needed expensive graphics terminals and slow communications limited the amount of user-interaction possible. Complex diagrams could take a significant period before appearing on the screen. Permanent output had to be redirected to a graph plotter or possibly to microfilm and users had to wait several hours before they obtained their output. The software available for mainframe graphics is either available as a library of subroutines such as GINO-F, GKS or TELL-A-GRAF or it may be included in a package which only gives the user limited control over his output. This situation still exists for many users.

Graphics software on microcomputers seems to present a far wider range of possibilities. This may be because part of the history of microcomputers has been associated with visually intricate computer games. For example most microcomputer programming languages have graphic-orientated routines. A typical example is BBC-BASIC, a language which enables the user to produce quite sophisticated coloured graphical output. It is interesting that many languages ported from mainframe machines need graphic enhancements when they are adapted to specific microcomputer use. In addition to the built-in graphic capabilities of a microcomputer it is possible to purchase graphic packages. These fall into two categories, those that accept data from existing files created by other packages or via an editing program, and those that are 'painting' programs where the computer screen is the canvas and a pointing device, usually a mouse, is the brush. In the first category are packages like the Harvard Graphics package and Lotus Freelance-Plus for IBM PC and compatible microcomputers and Cricket Graph for the Apple Macintosh. The latter is a delightfully flexible system that is simplicity itself to use. It allows data to be typed directly into the computer or read from files. It then offers a wide range of graph types which range from histograms to pie charts to standard x,y plots. All the attributes of the graph from symbol style and grid size and markings to font size and style for the caption are quickly and easily selected. Output from a laser printer is so clear that these graphs are of publication quality and constitute, to our minds, one of the most labour-saving tools that modern computer systems can supply to the academic and research community.

In addition to the flexibility and versatility of microcomputers with respect to graphics there is generally a greater variety of hard copy devices available to them. Thus monochrome output can be sent to a dot matrix or to a laser printer instantaneously. This can be a direct copy of the screen image or a scaled and interpolated representation of what appears on the screen. Desk-top plotters are also readily available. These can provide a wide range of pen colours, often six or eight, which are able to draw on paper or on acetate sheets. One can often interact with the plotter and change pens to obtain a greater range of colours or to use pens of different widths. This flexibility is not likely to be available with plotting devices connected to a central system.

It is often useful to generate 35 mm colour transparencies of material presented on the screen. There are two options open to the user. First direct photography of the screen itself. This is perfectly possible. Good results can be obtained if the user remembers to use long exposure times greater than about 1/8 of a second at an appropriate stop value for the film speed. The length of time ensures that the picture will not be affected by scan lines on the TV screen or monitor (Chapman, 10). Greater sophistication can be obtained at a higher price using the Polaroid Palette. This is a dedicated device with its own internal screen and colour filters.

4.2.6 *Integrated packages*

An integrated package, as its name implies, is one designed to contain all the software applications most commonly needed by the businessman. These are generally a word processor, a database management system, a spreadsheet and a communications package. A convenience factor underlying these packages is that the user interface is identical in each module so that the same commands are used as far as possible for each application. Even more important is that data can be shared between each module and an application can be developed using several of them. For example one might enter data into the database, transfer it to the spreadsheet where a graph is produced. The latter is then included in a report written on the word processor and sent electronically to a colleague using the communication package! Integrated packages may sometimes be criticised in that one of the modules is better than the others and it may be true that dedicated single function programs can be more flexible. However integrated packages may be cheaper than the cost of buying individual dedicated applications programs and in the long run they, or an operating system offering the same features, seem to be the way forward.

5. PROFESSIONAL COMPUTER INTEREST GROUPS IN MICROBIOLOGY

Microbiologists interested in applying computer techniques to their subject have access to a few specialist interest groups which are worth describing.

5.1 **The Society for General Microbiology Computer Users Group**

The SGM Computer Users Group was initially formed in 1983 as the SGM Computer Club, but is now a recognized interest group. It used to publish a 36-page newsletter (*Binary*) three times per year. *Binary* has since been taken over by Academic Press and will in future appear as a bimonthly news/journal. The SGM-CUG maintains a list of SGM members interested in computing and a program catalogue of microbiological software. The Group also organizes meetings at SGM symposia. For more information contact: Dr J.B.Evans or Dr J.W.T.Wimpenny, Department of Pure and Applied Biology, University of Wales, College of Cardiff, Cathays Park, Cardiff, UK.

5.2 The Microbiology Computer Users Group

This is a similar group organized by members of the American Society of Microbiology, although it was not until recently adopted by the ASM. It publishes an 8-page newsletter from time to time. The MCUG has just been adopted formally by the Board of Education and Training (BET) of the ASM. This rationalization step will give the group financial support with which to organize meetings and seminars. Contact: Dr R.M.Atlas, Department of Biology, University of Louisville, Louisville, KY 40292, USA.

5.3 The Public Health Laboratory Service Computer Users Group

The British PHLS organize a computer users group and publish a quarterly newsletter. The latter is devoted to computer applications in medicine and the health care area. Contact: Dr K.A.V.Cartwright, Public Health Laboratory, Gloucester Royal Hospital, Great Western Road, Gloucester GL1 3NN, UK.

6. JOURNALS, ETC. OF INTEREST TO MICROBIOLOGISTS USING COMPUTERS

6.1 Scientific publications

6.1.1 *Computer Applications in the Biological Sciences (CABIOS)*

CABIOS published by IRL Press, 'provides a forum for the exchange of information on the uses of computing in the biosciences'. It contains introductory, review articles, full papers, communications, application notes, a literature survey, product announcements, software, hardware and book reviews.

The editors are: Dr R.J.Beynon, Department of Biochemistry, University of Liverpool, PO Box 147, Liverpool L69 3BX, UK and Joseph L.Modelevsky, International Minerals and Chemical Corporation, 1810 Frontage Road, Northbrook, IL 60062, USA.

6.1.2 *Binary*

As mentioned above, *Binary* is the newsletter of the SGM Computer Club. From 1989 it will be published by Academic Press with Julian Wimpenny as Editor-in-chief. *Binary* is dedicated to covering applications of computing to all branches of microbiology. It will publish original refereed articles as well as continuing its tradition of a more informal approach to the subject. Thus it includes reports of meetings, general articles, review articles, news, helpful hints, letters and cartoons.

Contact the editor: J.W.T.Wimpenny, Department of Pure and Applied Biology, University of Wales, College of Cardiff, Cathays Park, Cardiff CF2 1XH, UK.

6.1.3 *Science Software*

This journal published by John Wiley, covers software that is of interest to scientists in general. It contains articles, full length and 'quick and dirty' reviews, a fleamarket, recent references, published programs and a features/departments.

The editor is: Dr Diana J.Gabaldon, Arizona State University, Centre for Environmental Studies, Tempe, AZ 85287, USA.

6.1.4 *Collegiate Microcomputer*

Collegiate Microcomputer was the brainchild of Brian J.Winkel; it publishes articles from all areas of higher education. Most are from the humanities but occasionally there are useful contributions from the life sciences. There have been a number of useful items on statistics and modelling packages which are very much worth reading.

Dr Brian J.Winkel can be contacted at: Rose-Hulman Institute of Technology, Terre Haute, IN 47803 USA.

6.1.5 *International Journal of Biomedical Computing*

This journal is produced by Elsevier and edited by J.Rose and J.H.Mitchell. It serves the medical sciences in particular but also covers some biological applications. Contact: Professor J.Rose, Department of Aeronautical and Mechanical Engineering, University of Salford, Salford M5 4WT, UK.

6.1.6 *Journal of Microcomputer Applications*

Published by Academic Press, this quarterly journal publishes interesting/unusual applications of microcomputers in any area of science and technology. Worth a glance. . . . . . .! Contact: M.J.Taylor, Department of Computer Science, University of Liverpool, PO Box 147, Liverpool L69 3BX, UK.

6.1.7 *Computer Methods and Programs in Biomedicine*

An Elsevier journal dedicated to biomedical topics that might be interesting to medical microbiologists and occasionally to other biologists. Contact: Professor W.Schneider, Uppsala University Data Centre, PO Box 2103, S-750 02 Uppsala, Sweden.

6.2 **Commercial magazines**

6.2.1 *Byte*

Byte is perhaps the most respected almost having the status of a scientific journal. It covers microcomputing of all types and has good in-depth articles and is often quoted in scientific literature.

Byte America, One Phoenix Mill Lane, Peterborough, New Hampshire, USA.
Byte Europe, McGraw-Hill Information Systems, McGraw-Hill House, Maidenhead, Berkshire SL6 2QL, UK.

6.2.2 *PC User and PC Week*

There are both UK and US versions of these publications. They are available free to those involved directly in computing (e.g. Computing Centres) but can also be purchased. They carry good reviews of all PC products.

6.2.3 *Personal Computer World*

PCW has an excellent coverage of hardware and software of interest to scientists and is certainly the best of the British monthly computer magazines. Current issues should be examined for the addresses of contacts.

6.3 **Miscellaneous titles**

6.3.1 *Agrenet News*

The newsletter of the British Agricultural and Food Research Council Computing Centre. Contact: AFRC Computing Centre, West Common, Harpenden, Herts, UK.

6.3.2 *The CTISS file*

Computers in Teaching Initiative Support Service. This newsletter is of interest to people particularly interested in applying computers to teaching. Contact: South West Universities Regional Computing Centre, University of Bath, Claverton Down, Bath, UK.

6.3.3 *Biotechnology Information News*

Biotechnology information is obviously interesting to many microbiologists. This slender newssheet may help. Contact: The British Library Biotechnology Information Service, Southampton Buildings, London WC2A 1AW, UK.

6.3.4 *Perspectives in Computing*

This journal is produced by IBM and must therefore be accepted in that light. . . . .! It is designed to cover computer applications in the academic, scientific and engineering communities. It is bright, cleanly produced and colourful. It has little that is directly relevant to microbiologists but it is free to academics and others who can write in on their own letter head. Contact: IBM Corporation, Armonk, NY 10504 USA.

7. SOURCES OF MORE GENERAL HELP

There are many Computer User Groups, whose names and address may be found in the telephone directory. The other source of information about these groups is to be found in the various computer magazines. Such groups are typically interested in microcomputers as opposed to mainframe, however mainframe-user groups do exist but are more likely to be composed of people running such systems.

PC-Sig, 1030D East Duane Avenue, Sunnyvale, California, USA.

PC User Groups USA, see phone book.

IBM PC User Group, PO Box 360, Harrow HA1 4LQ, UK.

COMPULINK User Group, Suite 2, The Sanctuary, Oakhill Grove, Surbiton, Surrey KT6 6DU, UK.

8. REFERENCES

1. British Computer Society (1985) *A Glossary of Computing Terms, An Introduction.* Cambridge University Press, Cambridge.
2. Lang,K. (1987) *The Writer's Guide to Desktop Publishing.* Academic Press, London and New York.
3. Bate,J.S. and Wilson-Davies,K. (1987) *Desktop Publishing.* Blackwell Scientific, Oxford.
4. McClelland,D. and Danuloff,C. (1988) *Desktop Publishing Type and Graphics.* Hardcourt, Brace Janvinovitch, Austin.
5. Martin,J.J. (1974) *Principles of Database Management.* Prentice-Hall, London.
6. Sharman,K. (1987) *Introduction to Database on Microcomputers.* Addison-Wesley, Wokingham, UK.
7. Kibby,M.R. (1985) *CABIOS*, **1**, 73.
8. Kibby,M.R. (1986) *CABIOS*, **2**, 1.
9. Kellis,J.T. (1986) *Backman Data Sheet,* **7773**, 41.
10. Chapman,S.J. (1984) *Binary*, **1**, 11.

CHAPTER 2

Image analysis

PETER J.H.JACKMAN

1. INTRODUCTION

Image analysis (IA) is an area of computer application in microbiology which should show considerable expansion over the next few years as the price/performance ratio of systems continues to improve. The range of image analysis systems is changing rapidly, however I shall attempt to give a review of concepts and methods together with a guide to systems currently on the market.

Since so many areas of experimental science involve human vision there are many applications for image analysis wherever repetition or quantification are required. However, with the exception of electron microscopy (EM), examples are still rather rare in microbiology. The general increase in awareness of, and access to, computing through the prevalence of the microcomputer is, however, now stimulating the development of applications. Electron microscopists were pioneers in the use of IA partly through the fact that it deals with images that can only be seen with the use of electronics and partly because the very high cost of EM hardware made the additional high costs of early image analysers more acceptable.

2. IMAGE ANALYSIS—FIRST PRINCIPLES

Images must first be converted into a form that can be manipulated by computers. This is achieved by digitization of the image into a grid of points or picture elements ('pixels') and the measurement of the intensity of light at each point. Black and white images may be recorded by a single measurement of intensity (grey level) and colour images by measurement of the intensity at three wavelengths (red, green, blue). A typical application might digitize an image into 512×512 pixels with an intensity scale of integer values 0–255. Such an image would occupy 256 kilobytes (K) of memory (*Figure 1*). Once an image is thus available in computer memory the pixel values may be subjected to any mathematical operations desired. It is important to realize that each operation will have to be repeated more than 256 000 times for each image so image analysis places great demands on computational speed and memory. Having acquired an image, typically it will need to be enhanced in some way perhaps to reduce noise, to increase contrast or to remove unwanted objects. In a second phase the desired features, for example bacterial cells, are separated from the rest of the image and measurements made, for example, of perimeter and area of the cells. It must be stressed that in most areas of image analysis the human eye/brain combination remains unsurpassed by computer programs. Computer analysis systems have advantages in terms of memory, quantitative measurements and in repetitive tasks. They are inferior in powers of resolution, discrimination, interpretation and response to unexpected phenomena.

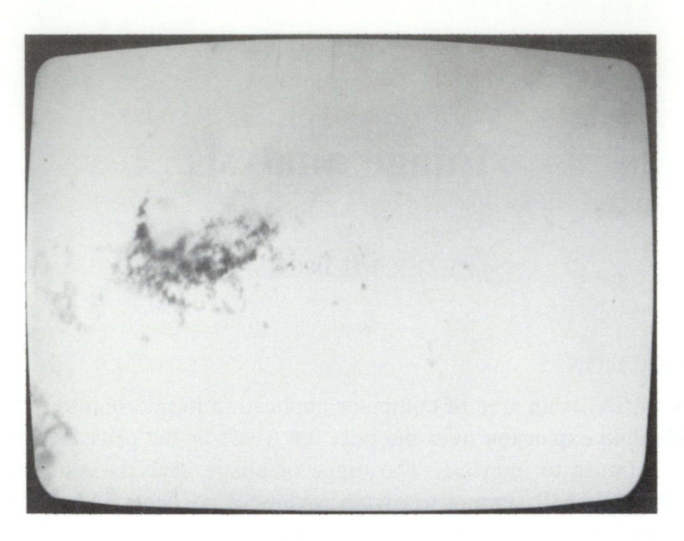

Figure 1. Image analysis of bacteria in food: digitized image.

2.1 Image input

A variety of image acquisition systems are available although most applications will be using a video camera. Choice of input device should be considered carefully and it should match the quality of the rest of the system.

2.1.1 *Digitizers*

Digitizers allow the user to move a stylus or cursor over the outline of the object to be studied, usually a photograph, and the position of the pen is transmitted as a stream of *x,y* coordinates to a computer. They usually offer quite high resolution, in excess of camera systems, and allow the user to interact with the analysis system at the input stage. Consequently the user is able to ensure that only the features of interest are transmitted to the analysis system and thus a variety of errors in the original image may be circumvented. The disadvantages are that only outline information can be collected and a good deal of user time is involved. Obviously human subjectivity plays an important part in the use of such a system; however tasks may be performed which would be beyond the capabilities of current IA systems. The most frequent use of digitizers is in collection of outlines from serial sections for three-dimensional reconstruction.

2.1.2 *Cameras*

The choice mainly lies between conventional tube-based cameras (Vidicon, Plumbicon, Newvicon and Chalnicon) and solid-state cameras (CCD, CID). The principal parameters of interest are light sensitivity, spectral response, signal-to-noise ratio and resolution. Vidicon cameras are the least sensitive, but they are robust to over-illumination and are suited to bright-field microscopy and many macroscopic uses. Chalnicons are the most sensitive and are suited to fluorescence applications. For ultra-low light levels a number of image intensifying devices are available. Light sensitivity is usually described as a minimum illumination level, typically 3–4 lux for a Vidicon, about 10-fold lower for a sensitive tube or solid-state camera, down to 10E-8 lux for an image intensifier. It should be noted that at the low end of its sensitivity range the signal-to-noise ratio of a camera will be far below its optimal level. Resolution varies from about 280 television lines for a colour CCD camera to 800 lines at the top of the Vidicon camera range. It is critical to test cameras for the desired application before purchase and not to rely on stated specifications. Resolution and signal-to-noise should be calculated and not visually assessed from the monitor picture. Solid-state cameras, for example charge-coupled device (CCD) cameras, are currently in a rapid state of development; they offer advantages of size, robustness, dynamic range and lack of geometric distortions. At present however they tend to have a lower resolution and worse signal-to-noise than tube cameras. Vidicons and Plumbicons have their optimum response in the blue–green whereas Newvicon and Chalnicons have a greater response in the red region of the spectrum.

2.1.3 *Densitometers*

Scanning densitometers may be used to digitize flat specimens or photographic negatives and transmit the image to a computer. They have the advantages of high spatial and grey level resolution and are without geometrical distortion; however they are slow, expensive and limit the kind of specimen which may be analysed.

3. DESIGNING YOUR OWN IMAGE ANALYSIS SYSTEM

Inevitably the best price/performance ratio is achieved by assembling one's own hardware and software rather than purchasing an off-the-shelf system. In considering this option the inexperienced should be very careful to assess the situation realistically. The advantages of the DIY approach include lowest initial cost and flexibility. However the cost of writing one's own software is very high if properly assessed and will invariably take much longer than envisaged. Programmers with IA experience are hard to attract into microbiology. Choice of operating system and language for IA may be restricted by the hardware. Considerations of speed would indicate C as the high level language of choice, however, programmers in BASIC, Pascal or FORTRAN are usually easier to find. The intermediate approach of buying a commercial package which has the capacity to accept additional user-written routines has much to recommend it. One may have to write additional routines in a particular language using a particular compiler in such cases. In addition, routines in a package may not be as fast as those that have been optimized for a specific application. Possibly the most effective strategy is to buy a package and to have any additional customization done by a contract programmer. In this case the scientist must have, at the outset, a precise idea of what the IA system should do.

3.1 **Hardware**

In general the computational demands of IA systems have necessitated the development of a range of dedicated hardware to give acceptable processing speeds. Recently attempts have been made to make use of general purpose microcomputers but in all systems some use is still made of specialized hardware in, for example, the frame grabber. In the future, development of general purpose parallel processors such as the Inmos transputer may allow the use of non-dedicated hardware for image analysis. Generally a trade-off has to be made between high-speed image processing in dedicated hardware and lower speed but more flexible processing by the general purpose computer systems.

3.1.1 *Proprietary designs*

The older generation of image analysers, such as Cambridge Instruments' Quantimet or Joyce-Loebl's Magiscan, use proprietary designs which naturally restrict user choice in configuration or source of hardware. The systems aim to offer a comprehensive package which provides a solution to an IA problem with the minimum amount of expertise demanded of the user. They are general purpose instruments but may have software packages, for example chromosome analysis or two-dimensional gel analysis, which can turn them into dedicated systems. For users who can afford the premium price of these sort of systems they frequently provide the ideal solution. A comprehensive discussion of the relative merits of this class of image analysers is beyond the scope of this chapter and readers are advised to take advantage of the enthusiasm of such companies to demonstrate their latest products.

3.1.2 *Bus-based systems*

A bus is the communications device linking components in an electronic system. Commonly designs based on IBM PC, VME or UNBUS systems are available where a number of a single board elements may be combined as required to build a system to a user's specific needs. There is also considerable scope for upgrading the system in the future. On the other hand there is usually a high initial cost or, in the case of the IBM bus, the bus itself may limit the speed. It should be noted that some systems only use the bus for convenience and may transfer images by another route. At the moment use of the Intel 80386 processor with the IBM bus or the Motorola 68020 with the VME bus would be considered as the best options available.
(i) *An IBM BUS system.* As an example the range of boards supplied by Data Translation Ltd for the IBM AT will be described. A fuller description of the functions mentioned can be found later in this chapter. The principal board is a 512 × 512 × 8-bit frame grabber (DT2851) which digitizes a video signal, stores an image in one of two onboard memory buffers and displays the image in RGB false colour or monochrome at rate of 30 images per sec. The frame buffers are memory-mapped directly into the extended memory space of the AT, leaving the lowest megabyte of AT system memory available to the user for image operations. Frame data from either buffer may be accessed at any time using standard 8-bit or 16-bit AT memory access operations. Real-time 8-bit arithmetic and logic image processing operations may be accomplished through the use of a look-up table (LUT). A LUT enables the results of arithmetic operations to be calculated in advance rather than during image processing. The LUT processor consists of a feedback path which allows a frame of data to be operated upon by one of eight 256 × 8-bit LUTs and returned to either of the frame stores and displayed, all in 1/30 sec. Operations include (i) averaging frames; (ii) adding and subtracting frames; (iii) multiplying and dividing frames by a constant; (iv) AND, OR, XOR logic operations; (v) contrast and brightness enhancement; (vi) windowing and region-of-interest processing; and (vii) graphic overlays.

IRIS analysis software includes functions for performing convolutions, averaging, histograms, windowing, arithmetic operations, logical operations, zoom, pan and scroll, graphic overlays, cursor display/manipulation. Commands may be grouped and stored as combination commands ('macros'). These functions may be speeded up by one or two orders of magnitude if they are implemented in hardware rather than software; for this reason an auxiliary frame processor (DT2858) is available. Performing convolution of an image with the board is 250 times faster than the IBM AT. Images are transferred between the frame grabber and the processor over a dedicated interface much more quickly than on the IBM bus. Further operations may be speeded up by using a floating-point array processor (DT7010). Such operations include digital filters, fast Fourier transforms and matrix operations. A library of subroutines for image processing using these boards called DT-IRIS is available. These may be called from one's own program in Microsoft BASIC, C, Pascal and FORTRAN.

3.1.3 *Array processors and parallel processing*

An array processor is a device which contains multiple processors able to carry out the same operation on different pieces of data simultaneously. This activity is typical of IA and hence an array processor is a very desirable part of an IA system. Examples include the CLIP, DAP, Imagine and Sky processors. Unfortunately the cost of such devices to date has been high. Array processing is one type of parallel processing; however, in an advanced system, it ought to be possible to perform different operations on multiple data simultaneously. Systems based on arrays of the Inmos transputer are capable of this and are now commercially available. In discussion of image processing the expression 'pipelining' may be encountered. This means that while one part of the processor is performing an operation, another part is accessing memory access so that the processor no longer has to wait for the next piece of information.

3.1.4 *Transputers*

The Inmos transputer is a processor designed from the outset for parallel processing. It is now being introduced into both large scale dedicated parallel computers or as add-on units in IBM or VME bus-based systems. Such systems are ideal for image processing and their introduction is considered important enough to describe here ahead of their application in microbiology. Each transputer has four input/output channels called 'links' that operate at very high speed (20 Mbits/sec) and simultaneously with program execution. The links enable transputers to be connected into arrays like children's building blocks. The arrays may be of any size and many different topologies. Overall performance is directly proportional to the number of transputers used. A parallel processing language called OCCAM has been designed to program transputers efficiently, and programs may be developed on a single transputer and then executed on an array without further alteration to the program. A supercomputer based on transputers has been designed by Meiko Ltd. One transputer is dedicated as a 'logical host' performing program development, housekeeping tasks and communication to an external computer which may be used for file storage. Another transputer is dedicated to display graphics and another as a hard disc interface. The main computing power is supplied by 'compute' elements each containing four transputers. Up to 40 compute elements may be installed to produce a computer capable of 1175 MIPS (million instructions per second) for a cost of about £250 000. At the other end of the scale single transputer boards together with the OCCAM language for the IBM bus are available for about £4000. The latest transputer, the T800, includes a floating point unit performing 64-bit floating point arithmetic at a rate of 1.5 MFLOPS (million floating point operations per second). A 20 MHz T800 transputer performs about 10 times faster than an Intel 80286/80287 combination and five times faster than a Motorola 68020/68881 judged by the Whetstone Benchmark.

3.2 **Software**

Most commercial software systems provide a menu of image processing operations which the user can select to successively acquire, process and make measurements of an image. Some systems allow the recording of a series of these operations to form a high level image processing language. As a further extension of flexibility it may be possible to code one's own algorithm in a language such as FORTRAN or Pascal and to add these to the image processing language. Systems with the latter facilities include the Joyce-Loebl Magiscan and the Synoptics Semper software. In general, commercial systems provide a rapid way of experimenting with image processing functions and of course require little or no knowledge of programming.

4. IMAGE PROCESSING—SOME TYPICAL OPERATIONS

4.1 **Shading correction**

The image captured by a video camera will usually have a far from uniform background illumination partly due to shortcomings in the camera and partly from uneveness in specimen illumination. This shading may not be obvious to the eye but becomes apparent when thresholding is attempted. A background image should be smoothed and then used to correct specimen images by subtraction or multiplication.

4.2 **Contrast enhancement**

One method of enhancing the contrast of an image is histogram equalization (*Figure 2*). A histogram of a number of pixels at each grey level will show the majority of the pixels to occupy a limited range of all the possible grey levels. If the distribution of pixels is re-calculated to be spread over a wider range of grey levels, then the contrast of the image will be improved. It should be noted that this type of process only improves the visual appearance of an image and does not assist in further quantitative analysis.

4.3 **Convolution**

This term covers a variety of operations on the central pixel of a matrix of pixels based on the values of its neighbours. The matrix might be a square of nine pixels or more. Examples include: spatial averaging where the value of the central pixel is made equal to the average of its neighbours; median filtering where the intensity of the central pixel is re-calculated to be the average of or equal to its neighbours; or min–max where the maximum and minimum neighbouring values are found in the matrix. If they differ by more than a set threshold the central pixel is replaced by whichever of these is nearest its value. Otherwise it is replaced by the average of its eight nearest neighbours. This has the effect of sharpening edges and smoothing the remainder of the image.

Figure 2. Contrast stretched by histogram equilization.

Convolution is computationally intensive and for that reason some systems carry it out in hardware.

4.4 **Edge finding**

Edges may be found by measuring the gradient along a line of pixels and their rate of change; edges occur where the rate of change of gradient crosses zero. Edges may be found by convolution along the line with the Laplacian or Sobel operators.

4.5 **Fourier analysis**

Fourier analysis transforms the spatial information in an image into frequency information. For example, noise consisting of small rapid changes of gradient gives rise to high frequency components which may then be removed.

4.6 **Signal averaging**

If multiple images are summated and then averaged, random noise will be smoothed out. This is particularly useful with faint noisy images.

4.7 **Erosion and dilation**

Removal of pixels around the edge of objects (erosion) increases their separation, whereas adding pixels around the edge (dilation) merges them. Erosion to a single line is called skeletonization.

4.8 **Boundary detection**

When an object has been detected, for example by thresholding (*Figure 3*), it is then necessary to store its boundary in order to be able to count objects in a field of view or apply further processing to the objects only. One method is by chain coding. A

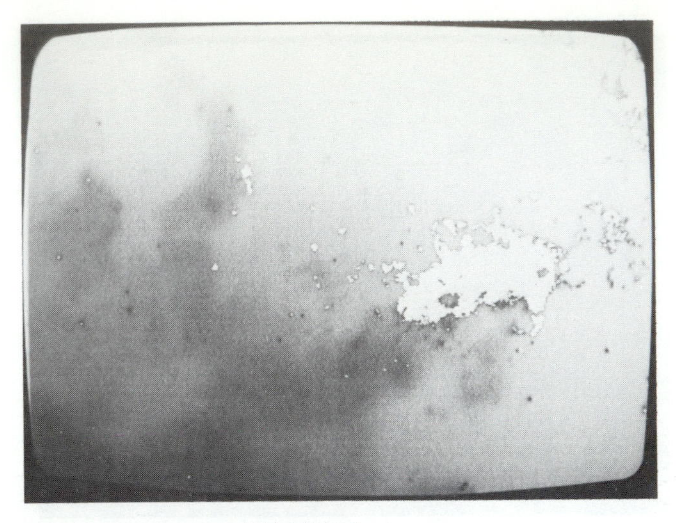

Figure 3. Region darker than a threshold highlighted.

program to do this would start at the bottom left of the field of view and search until the first pixel of the object is found. The program then examines the eight pixels around that pixel in a clockwise direction until it finds the next connected pixel. This is repeated until the starting point is reached again. Just the direction of the next pixel at each point is stored and this needs much less memory than storing the (x,y) position of each boundary pixel.

4.9 **Feret diameter**

This measurement may be used for particle sizing. The horizontal and vertical diameters may be measured by finding the maximum and minimum x and y coordinates.

4.10 **Shape factors**

These are measurements which can be used to classify objects by shape regardless of orientation. An example is circularity.

$$4\pi A/P^2$$

where A is the area and P the perimeter length of an object. This expression has a value of 1 for a circle and decreases with increasing departure from circularity.

4.11 **Colour**

The use of colour in IA is still at a preliminary stage due to the high hardware costs. Colour has three parameters: intensity, hue (which is the wavelength of maximum apparent intensity) and saturation which is the purity of the hue from other wavelengths. Colour television encodes the red, green and blue elements of a signal. Monitors to display such images are consequently termed RGB. Colour cameras may have three tubes each filtered to respond to red, green and blue or use a single detector with a filter with microscopic regions for each colour. It takes three times the memory to store the three hues compared with monochrome images; however this may be reduced in some systems, for example to three 4-bit components (4096 possible colours) rather than three 8-bit components. Colour image analysis allows the detection and classification of objects which have a similar grey level, size and shape but different colours.

5. AREAS OF APPLICATION IN MICROBIOLOGY

The main area of microbiology where IA is used to any extent at present is in the counting of colonies growing on media in Petri dishes. Other areas just beginning to be developed include: counting fluorescent cells on filters in the DEFT technique, detection of colonies in sections by automated light microscopy and some applications in laboratory measurement. In addition, of course, there are well developed applications in electron microscopy.

5.1 **Direct epifluorescent filter technique (DEFT)**

In microbial enumeration using the DEFT system bacteria are collected on a filter and stained with acridine orange. The cells are then counted under an epifluorescent microscope. Pettipher and Rodrigues (1) used a Chalnicon camera and semi-dedicated IA system (Optomax System III, Micromeasurements Ltd, Shirehill Industrial Estate, Saffron Walden, Essex, UK) to count fluorescing cells or clumps of cells. Simple thresholding was used to detect cells (*Figure 4*). Problems found included out of focus cells at the periphery of the field, some poorly stained bacteria and occasional somatic cell nuclei and debris. In general, however, the counts were found useful and showed no more higher variation than did the standard visual counting technique.

5.2 **Quantification of growth**

IA has been used by Caldwell (2,3) to quantify microbial growth in slide cultures. Video images were collected at 15-min intervals during incubation, stored in RAM and aligned with a reference image collected at time 0. A shading correction was used to produce a uniform illumination and remove any lens debris. A median filter was then used to remove aberrant pixels from the image. The time 0 image was subtracted from each of the subsequent images. The resulting difference images were normalized and then filtered to enhance images. This resulted in an image with black objects which represented biomass formed since time 0. As the cells grew they often shifted slightly giving a grey and white image of the cell as it appeared initially, and a grey and black image of the cell as it appeared after incubation. The grey area was the overlap between the initial and final images. The white area was the area vacated owing to movement of the cell away from its initial position. The total white area (cell movement) was subtracted from the total black area (cell movement plus cell growth) to determine the increase in area which was due only to growth. Caldwell suggests that this difference image technique could also be applied to the study of metabolic activity of cells with fluorescent or enzyme stains.

5.3 **Planktonic bacteria**

A number of workers have used image analysis to measure bacterial numbers in sea water. The experimental approach is more or less the same as with DEFT, that is a volume of water is filtered through a membrane and the cells stained with acridine orange

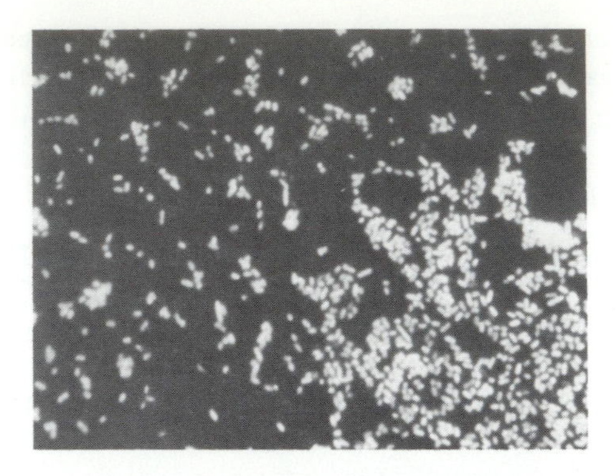

Figure 4. Digitized image of bacteria stained with acridine orange for DEFT.

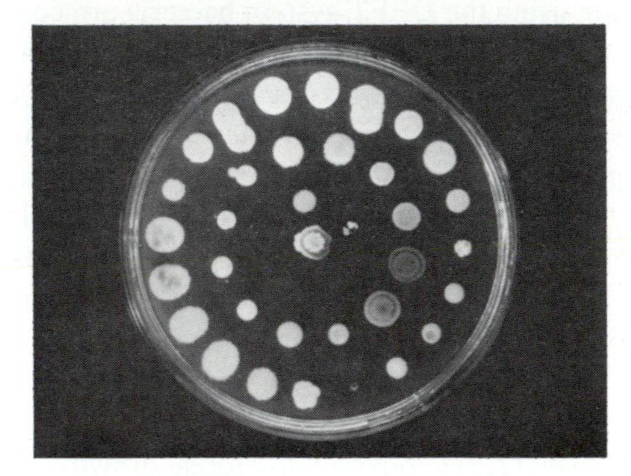

Figure 5. Digitized image of a multipoint inoculated plate.

or the DNA stain 4′,6-diamidino-2-phenylindole (DAPI). Sieracki *et al.* (4) used a dedicated system (Artek 810, Artek Systems Corp., Farmingham, NY, USA). Both edge detection and thresholding algorithms were used. A light pen was employed to edit out extraneous objects. A variety of measurements on each object in the image could be made, including perimeter, area, longest dimension, longest horizontal chord, horizontal and vertical ferets, circularity and location coordinates. Their scale factor was 0.20 μm per pixel with a $\times 100$ objective. The majority of bacterial silhouettes contained 7–20 pixels. Measurement of biovolume is thus not very accurate for small cells but measurements in general were found to be as accurate as visual methods.

5.4 Multipoint inoculated plate reading

Breakpoint antibiotic sensitivity testing has been semi-automated by using IA to measure bacterial growth on a control plate and on a range of test plates (*Figure 5*). In the same way minimum inhibitory concentrations (MICs) and screening urines for bacterial growth may be performed.

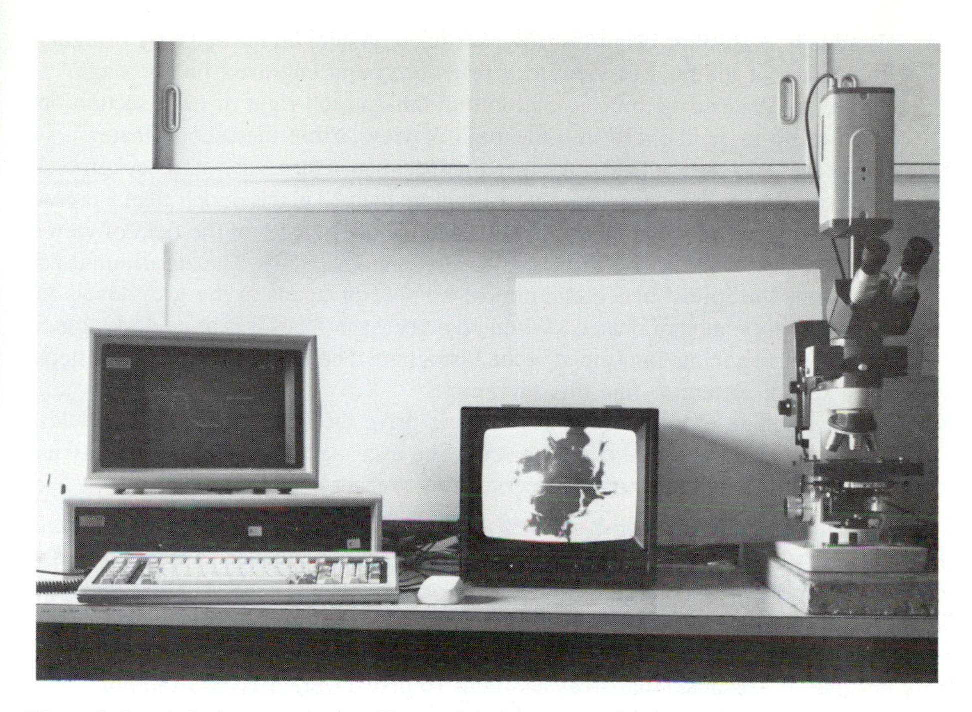

Figure 6. A typical microcomputer-based image analysis system: an IBM XT compatible with frame grabber board, 512 × 512 monitor and video camera mounted on microscope.

6. CASE STUDY—AUTOMATED DETECTION OF MICROORGANISMS IN FOOD

In the author's laboratory an IA system was developed to automate as far as possible the detection and quantification of microbial growth in sections of food (5). With a limited hardware budget a system based on an 8086/8087-based IBM PC (Compaq Deskpro, 640 K RAM, 10 Mb hard disc) and a single card with frame grabber, frame store and LUT functions (PC Vision, Data Translation Ltd) was selected. This system is now some 2 years old and considerably more powerful hardware such as Intel 80386 based system can now be purchased for a similar cost. The system acquires its images from an inexpensive black and white video camera (Hitachi) mounted on a medium quality light microscope (Richert Jung Microstar One-Ten). A motorized 12 inch × 3 inch stage was custom built for the microscope (Apollo Optical Ltd). The stage controller (McClennan Servo Systems Ltd) powers and controls three stepper motors, two driving the stage and one driving the microscope fine focus control. The stage controller accepts commands from the computer over an RS232 interface (*Figure 6*).

A program to detect Gram-positive bacteria in sections of heterogeneous foods was written in Turbo Pascal (Borland International) and used the following steps.

(i) Acquire an image in the absence of a specimen and store. Subtract this reference image from specimen images to remove the effect of non-uniform illumination emanating from the microscope lamp and condenser system.

(ii) Establish an absolute origin for the coordinate system on the stage by bringing the centre of the field of view to a reference point engraved on the stage.
(iii) Drive the stage so as to bring the bottom left and top right of each section on each of up to eight slides into the field of view. Store these coordinates.
(iv) Drive the stage to each field of view in the specimen in turn, subtract a background image and then, in fields where there are more than 10 pixels darker than a preset grey level, record the number of pixels and the coordinates of the field of view.
(v) Every 10 fields the microscope is re-focused automatically. The algorithm used measures the spread of a histogram of number of pixels in the grey levels in the intensity region of objects. When an object is in focus it is at its darkest and there will be a minimum spread in the histogram. The objective is driven in steps in a search pattern to find this minimum.
(vi) After scanning of all sections is complete drive the stage to display all fields of view containing thresholded objects. The user may confirm their identity as bacterial or as other stained objects which are not bacteria.

In the foods under investigation bacteria are few and far between and so the system described here has been successful in greatly reducing the number of fields of view which the operator needs to examine personally. In a run of 3122 fields of our material, 521 fields (16.7%) were recorded as positive using a conservative threshold level which minimized false negatives. Human examination showed that 18.4% of these fields were truly positive. It was noted that areas less than 10 pixels were derived from noise and when these were eliminated positives were about 3% of total fields of which about one-third were true positives. Thus although the system is still dependent on humans to discriminate between bacteria and other stained material the amount of human observation required is reduced 30-fold.

Attempts to develop a program to detect colony boundaries and thus to count colonies automatically revealed a significant level of subjectivity in exactly what were colony boundaries according to human observers. It became apparent that, in contrast to growth on artificial media, colonies in food may be fragmented and do not necessarily consist of cells in contact with one another. It was therefore decided to measure total bacterial area/field of view rather than number of colonies. The discrimination of bacteria in our material is impossible to make by grey level, size or shape alone and parameters such as texture and colour may be needed to raise discrimination to a level adequate for complete automation.

The 8086/8087-based IA system described here was constructed to a limited budget and the processing speed of this class of system should be considered where the number of images to be analysed is significant.

Currently the system takes about 8 sec to process each field of view. Increasing throughput can only come about by employing a faster host computer, for example machines based on the Intel 80386 processor with dedicated image processing boards and array processors. At some point the number of slides which may be loaded onto the stage will become limiting and some form of automated slide loading from a large capacity cassette will be needed.

The fully automated detection of microorganisms in natural environments is a challenging task even for the most sophisticated IA system and difficulties should not be underestimated. Our work has indicated that it is possible, but that much more work is needed to produce a completely effective system.

7. CONCLUSIONS

Undoubtedly the applications of IA to microbiology will increase as the price/performance ratio of systems increases. Any aspect of microbiology involving microscopy is now open to automation and studies of microorganisms *in situ* in a whole range of natural environments such as water, soil, food and clinical specimens should soon benefit. Applications to invasion of plant leaves by pathogens and studies of biofouling films are known to be underway. General uses around the laboratory include, measurement of colonies and other effects on artificial media including growth on gradient plates, analysis of protein and DNA separations in electrophoretic gels and measurement of many test results made visually at present.

8. ACKNOWLEDGEMENTS

Paul Plover of Analytical Measuring Instruments Ltd is thanked for provision of *Figures 4* and *5*. Mark Fernandes is thanked for provision of *Figures 1, 2, 3* and *6* and advice.

9. REFERENCES

1. Pettipher,G.L. and Rodrigues,U.B. (1982) *J. Appl. Bacteriol*, **53**, 323.
2. Caldwell,D.E. and Germida,J.J. (1984) *Can. J. Microbiol.*, **31**, 35.
3. Caldwell,D.E.J. (1985) *Microbiol. Methods*, **4**, 117.
4. Sieracki,M.E., Johnson,P.W. and Sieburth,J.M. (1985) *Appl. Environ. Microbiol.*, **49**, 799.
5. Fernandes,M.A., Jackman,P.J.H., Clark,S.A. and Gunard,S.R. (1988) *CABIOS*, **4**, 291.

10. FURTHER READING

Image Analysis: Principles & Practice (1985). Published by Joyce Loebl, Gateshead, UK.

11. APPENDIX

Systems checklist

System name: Visage
Supplier: BioImage Corporation, 1460 Eisenhower Place, Ann Arbour, MI 48108−9990, USA
Host computer: MassComp 68020 1. 5 Mb 50 Mb 1 Mb Multibus UNIX/C/FORTRAN
Image acquisition: flat bed scanner with 1728 element camera
Display: 1024 × 1280
Resolution: 1024 × 1024 × 256
Software: spot detection and measurement, image alignment
User routines: yes: UNIX/C/FORTRAN
Special applications: analysis of two-dimensional electrophoresis gels
Comments: colour option

System name: IPS/68K
Supplier: Kontron Electronics Ltd, Blackmoor Lane, Croxley Centre, Watford, Herts WD1 8XQ, UK
Host computer: custom 68000 1 Mb 40 Mb 600 K, Multibus
Dedicated hardware: pipeline processor, real time processor
Resolution: up to 4096 × 4096
Display: 1280 × 1024
Software: image processing library, database, statistics
Special applications: IBAS 2000—light microscopy, SEM-IPS—scanning electron microscopy, MIPRON—digital subtraction angiography, cardiology, nuclear medicine, IPS Ophthalmic. DEC-IPS for DEC computers
User routines: yes/C/Pascal/FORTRAN 77

System name: Optomax V
Supplier: AMS, Analytical Measuring Systems, Shirehill, Saffron Walden, Essex CB11 3AQ, UK
Host computer: dedicated processor and IBM PC
Software: functions include area, feature count, end count, vertical intercept, horizontal intercept, perimeter, frame area, area fraction, count/unit area, intercept/unit area, mean distance between boundaries and ASTM grain size, X-feret, Y-feret, form factor, histograms of feature measurements
Comments: automatic stage and focus available

System name: 40–10
Supplier: AMS, Analytical Measuring Systems, Shirehill, Saffron Walden, Essex CB11 3AQ, UK
Host computer: dedicated processor
Image resolution: 384 × 280
Display resolution: 384 × 280
Special applications: colony counting, DEFT, inhibition zone diameters, spiral plate counting, MIC/break-point determination
User routines: no
Comments: automatic stage and focus available

System name: VIDS III
Supplier: AMS, Analytical Measuring Systems, Shirehill, Saffron Walden, Essex CB11 3AQ, UK
Host computer: IBM PC
Comments: interactive system with graphics tablet and camera. Colour input available

System name: Microtex 8700
Supplier: Microtex scientific Imaging Systems, 80 Trowbridge St, Cambridge, MA 02138, USA
Host computer: DEC MicroVAX II, hard discs to 474 Mb
Image acquisition: solid-state, laser scanners, SEM
Display: 280 × 1024
Dedicated hardware: display processor
Software: Microtex Imagenet
User routines: yes
Comments: communications with other DEC systems

System name: Intellect 100
Supplier: Micro Consultants Ltd, Kenley House, Kenley Lane, Kenley, Surrey CR2 5YR, UK
Computer: DEC LSI 11/23
Resolution: 512 × 512
Software: KIAS, ISPEC
User routines: yes: FORTRAN
Real-time analyser processor available

System name: IMAGINE
Supplier: Synoptics Ltd, 15 The Inovation Centre, Cambridge Science Park, Milton Rd, Cambridge CB4 4BH, UK
Host computer: options include Torch XXX. Sun, VAX, IBM PC and compatibles
Dedicated hardware: framestore, array processor
Functions: real time frame processing
Software: Semper
User routines: yes

System name: VIP
Supplier: Sight Systems, PO Box 37, Newbury, Berkshire, UK
Host computer: BBC or IBM PC
Image resolution: 256 × 256 or 512 × 512
Software: general package
Comments: low cost

System name: I3000
Supplier: Seescan Ltd, Unit 9, Gwydir Street, Cambridge, UK
Host computer: dedicated
Acquisition: 256 × 256
Software: Library included in BASIC
Special applications: colony counting
User routines: yes

System name: Quantimet 520
Supplier: Cambridge Instruments, Viking Way, Bar Hill, Cambridge CB3 8EL, UK
Host computer: IBM PC
Image resolution: 512 × 512
Software: included
Comments: automatic storage and focus options

System name: Quantimet 970
Supplier: Cambridge Instruments, Viking Way, Bar Hill, Cambridge CB3 8EL, UK
Host computer: DEC LSI-11/73
Image resolution: 896 × 704
Software: QUIPS
Special applications: autoradiography, fibre analysis, metallurgy.

System name: Magiscan 2
Supplier: Joyce-Loebl Marquisway, Team Valley, Gateshead NE11 OQW, UK
Host computer: dedicated
Image resolution: 512 × 512
Software: GENIAS user options in Pascal
Special applications: metallurgy, industrial inspection, chromosome analysis, colour

System name: micro-Magiscan
Supplier: Joyce-Loebl Marquisway, Team Valley, Gateshead NE11 OQW, UK
Host computer: IBM-PC/XT
Image resolution: 256 × 256
Software: included, user options: no

Board products

System name: PCVISION
Manufacturer: Imaging Technology Ltd
Supplier: Amplicon Electronics Ltd, Richmond Rd, Brighton, E.Sussex, UK
Host computer or bus system: IBM PC
Functions: frame grabber, store
Image resolution: 512 × 512
Software: ImageAction package, Itex/PC library
User routines: yes: C, Pascal, FORTRAN

System name: Series 100
Manufacturer: Imaging Technology Ltd
Supplier: Amplicon Electronics Ltd, Richmond Rd, Brighton, E.Sussex, UK
Host computer or bus system: IBM AT, VMEbus, Multibus, Q-BusPC
Functions: frame grabber, store, real time processing including image subtraction, averaging, arithmetic and logical operations
Image resolution: 512 × 512
Software: Itex 100 library
User routines: yes; C, Pascal, FORTRAN

System name: Series 150
Manufacturers: Imaging Technology Ltd
Supplier: Amplicon Electronics Ltd, Richmond Rd, Brighton, E.Sussex, UK
Host computer or bus system: VMEbus
Functions: modular cards, analogue/digital interface, frame buffer, pipeline processor, real-time convolver
Resolution: 512 × 512
Software: Toolbox × 150
User routines: yes

System name: DT2851
Manufacturer: Data Translation Ltd, The Business Centre, Wokingham, Berks RG11 2QZ, UK
Host computer or bus system: IBM AT
Functions: frame grabber, store, image processing
Image resolution: 512 × 512

Software: Iris library
User routines: yes; C, Pascal, FORTRAN
Comments: auxiliary frame processor DT2858 and array processor DT7010 available for image processing in hardware

Software products

Name: SEMPER V
Supplier: Synoptics Ltd, 15, The Inovation Centre, Cambridge Science Park, Milton Rd, Cambridge CB4 4BH, UK

SEMPER is a library of image processing routines written in FORTRAN.
Functions include.

(i) Arithmetic operations on pictures: subregion extraction/insertion stretching and/or rotation.
(ii) Calculation of histograms and grey level statistics: linear and arbitrary non-linear grey level matching.
(iii) Picture display in grey-level, contour, *y*-modulation, overprinted and line-graph forms, display lettering, use of a spot/crosswire cursor on the display screen.
(iv) Fourier, Walsh and Hilbert transformation—the use of complex data means that Fourier transforms can be treated as flexibly as other pictures.
(v) Filtering using transforms or direct convolution (for smoothing, differentiation, background levelling, de-blurring and aberration correction.
(vi) Diffractogram and auto-correlation function calculation; cross-correlation function calculation; mutual picture registration in position and angle.
(vii) Rotational, linear and lattice averaging; projection and back-projection user routines may be added in FORTRAN, SEMPER may be run on most computers under UNIX or on the IPM PC (Micro-Semper).

CHAPTER 3

Data analysis in microbiology

N.BRATCHELL, A.G.O'DONNELL and H.J.H.MACFIE

1. INTRODUCTION

There is a growing trend towards laboratory automation and greater use of analytical instrumentation. The modern microbiology laboratory accommodates such instruments as gas chromatographs, mass spectrometers, high-performance liquid chromatographs, UV spectrophotometers and instruments which can monitor bacterial counts on large numbers of samples very simply and rapidly using principles of conductivity and light absorption. A feature of such instruments is that they generate data rapidly, often comprising many measurements on individual samples. The task facing microbiologists is to convert this experimental data into meaningful information about the process or system under investigation. Data analysis is used to accomplish this but, by contrast with previous problems, pencil and paper can no longer cope with the mass of data generated by modern instruments. Instead the microbiologist requires computer-based analysis systems.

This chapter is intended to help microbiologists identify appropriate data analytic techniques and subsequently choose suitable computer packages or programs. There are many statistical techniques which often provide different, though complementary, solutions to the same problem. These have been divided in a traditional way into analysis of variance, regression and modelling, optimization and multivariate analyses. Many other subjects have not been covered because they are covered more thoroughly in other chapters, but they have been mentioned here to emphasize the need to consider packages in the light of all of the (widely) applicable techniques. There are also many other aspects to be considered when choosing a package such as data manipulation and storage, graphics, back-up support and user-friendliness.

2. THE PACKAGES

There are many computing packages aimed at a wide variety of markets for both main-frame, mini and microcomputers. They can be crudely divided into three categories: purpose-written programs, menu-driven packages and command-driven or programming languages. This section will give details of some of the most widely available general-purpose packages; more specific programs and packages will be introduced in later sections. There is also a vast number of other, often specifically-directed, programs available which will not be listed since it is likey that such a list would be out of date prior to publication of this chapter. The National Computing Centre maintains a directory of software packages for microcomputers, *Binary*, the Newsletter of the Society of General Microbiology Computer Club, regularly publishes details and reviews of new programs, and the Study Group of Computers in Survey Analysis publishes an extensive list of statistical software. Publications such as those of Wetherill and Curran (1), Berk (2) and Yeo (3) provide many useful comments on choosing software, and Crunch Software Corporation publishes a brochure giving useful tips on how to approach software reviews and vendors. A list of addresses for these publications and the various packages is provided at the end of this chapter.

High-level programming languages such as FORTRAN, ALGOL and Pascal are useful for programming novel techniques or routine analyses in the absence of more sophisticated packages. Their usefulness is augmented by the inexpensive library of NAG routines (Numerical Algorithms Group Ltd). The statistical routines range from simple statistical calculations, through analysis of variance, to non-parametric statistics; other routines cover problems in numerical analysis and matrix algebra. Although written to fulfil requirements of speed, accuracy and reliability, they require experienced programmers and are not suitable for the novice. Somewhat similar to the NAG routines is BMDP, a series of statistical routines with common input and output formats written for medical scientists. In addition to standard library routines, various books (4–6) and statistical journals (such as the *Journal of the Royal Statistical Society, Series C*; and the *Statistician*) contain programs and routines that may be used directly. As with the library routines, a high level of reliability can be assumed (with the above publications) but reliability does not guarantee that the results are sensible—that depends solely on the use to which they are put. A recent book by Thisted (7) may prove valuable to readers, in particular those who are interested in writing their own programs.

There are several well-established statistical programming languages, GENSTAT, MINITAB, SAS and SPSS-X. These were written for mainframe and minicomputers and can be run either interactively or in batch. Recent years have seen the development of microcomputer versions of these packages offering most of the facilities of their mainframe versions. They are similar in appearance in that they operate through a series of commands, like a programming language, but a command may be a simple function, such as a calculation, or it may invoke a very complex analysis, such as regression. In general they provide very flexible input/output facilities, a wide range of data manipulations and calculations, low- and high-quality graphics and macro or procedure facilities as well as a full range of statistical analyses. Their differences lie in ease of use, flexibility and emphasis.

(i) MINITAB is the simplest to use, but the least flexible as it was originally designed as a teaching package. It retains this role but now permits very detailed and complex analyses.
(ii) GENSTAT is perhaps the cheapest and arguably the most powerful and versatile package. Historically it has been regarded as very unfriendly and poorly documented and very much a statistician's package. It still has the latter quality but in its latest release (version 5) is more accessible to other users.
(iii) SAS is the most expensive and ultimately the most comprehensive package. In its basic form SAS is user-friendly but limited; extra flexibility and enhancements are purchased as extra modules.
(iv) Historically SPSS-X was aimed at the social sciences with emphasis on analysis of survey data, but it is now being aimed at industry and commerce as 'Serious Data Analysis Software'.

All of these languages possess procedure or macro facilities and other programming features which allow user-friendly forms of standard analyses to be programmed and retrieved in the same environment as novel or non-standard techniques. GENSTAT is provided with a library of procedures written by users covering less commonly used techniques, and SAS and MINITAB publish procedures written by users.

STATGRAPHICS is a menu-driven statistical package for the IBM PC. It is particularly useful for novices and occasional users as procedures are chosen from a menu which results in a procedure input panel being displayed, providing spaces for necessary parameters to be entered. The analyses offered cover the range of statistics. There are extensive file-handling procedures, and both graphical and numerical output can be edited and stored or printed. For users who require novel or unusual forms of analyses, STATGRAPHICS contains facilities for running user-written procedures, incorporating, if necessary, standard STATGRAPHICS routines, and procedure input panels. This package is the best example of a menu-driven package for microcomputers. Most other packages are less comprehensive, but the statistical packages listed above are now available for microcomputers and have an excellent pedigree.

The concepts of a mouse, pull-down menu and window displays have been thoroughly established by the Apple Macintosh and are becoming available for other systems. CRICKET GRAPH and STATVIEW both combine these features with a spreadsheet format for use on the Apple Macintosh. CRICKET GRAPH is a data display package offering various scatter plots and charts, whereas STATVIEW is a more general statistical package which provides graphics, data management, calculations and transformations, and a selection of analyses such as regression, analysis of variance and multivariate analyses.

3. PERIPHERAL CONSIDERATIONS

This section considers several peripheral aspects of computer packages. These are often overlooked, but are nevertheless important.

3.1 Some important questions

To choose the correct type of system the following questions must first be answered.

(i) What computing facilities are currently available? In particular, are they micro-computers, mini or mainframe computers? The choice of package or system has in the past been strongly affected by the answer to these questions. With the growing power of microcomputers and availability of hard discs and networks, the distinction is becoming less important, but it does still exist. At large institutions it is often cheaper to load mainframe packages than give each user a microcomputer version of the same package.

(ii) What statistical techniques are required? It is advisable to make a complete list. Packages do not all offer a full range of techniques, but it may also be that techniques are not available in a familiar form. It is also likely that the more esoteric or novel techniques are available only as purpose-written programs or may even need to be programmed.

An important point to consider is whether a system covers all of the necessary techniques or facilities. Transferring data between systems remains time consuming, even when it is possible. Ideally all operations should be carried out using a single package. Since it is still not always possible to transfer data easily from scientific instruments, these comments also apply to 'on-line' systems. Moreover there are problems of familiarity to be considered when using more than one system.

(iii) Who is going to use the package, and how often? There are numerous types of user: ranging from those with only an occasional need to those who will use the package frequently. Some will require simple standard analyses to be performed routinely and will need little flexibility; whilst others may need unusual analyses and great deal of flexibility. So is the system simple and is it flexible enough? Does it meet the various needs of different types of users? Packages are used by many diverse people and it is unlikely that all users can be satisfied by a single package. However, the choice will be moderated by the availability of expert advice or of programmers, who can make a relatively unfriendly package easier to use.

(iv) Who will be the local adviser? This matter should be considered carefully. It is surprising how much help may be needed even with the friendliest of programs. Users quickly confuse expertise in using one program with an all embracing knowledge of both statistics and computing! Providing advice is often enjoyable and challenging, but it can also be time-consuming.

(v) Is there good documentation and program support? Poor documentation wastes time and causes a good deal of frustration. Who provides the answers when neither the manuals nor the local adviser suffice? What happens when the program contains bugs? How reliable is the supplier or author of the software?

(vi) What is the integrity of the package? There are two parts to this problem. The first concerns the type of analysis offered to tackle particular problems or types of data. An incorrect or inappropriate analysis can lead to seriously misleading conclusions. Integrity also concerns the numerical routines used for the calculations. A routine which does not have high accuracy, for instance due to rounding errors, or is not otherwise numerically sound will give erroneous results, which may give rise to misleading conclusions. Programs and packages listed throughout this chapter have good back-up support and their integrity can be assumed.

(vii) What will be required in 1 or 2 years' time? Can the existing package be run on new hardware? Will the package be developed by the authors in response to changing analyses and requirements? Or can it be enhanced independently? Changing packages can mean a lengthy process converting all the data files to the new format and learning the idiosyncrasies of the package. Another consideration is the cost of upgrading software.

Answering these questions provides a basis for selecting the correct system, but there are further points to consider.

3.2 *Modus operandi*

Quite apart from the facilities offered by packages and programs, is the question of how they are actually operated or run. For example are they run in batch or interactively?

Thus background batch facilities are not available on microcomputers and there is a growing trend with minicomputers towards interactive real-time computing, although for very large data sets and complex analyses batch operation is still very useful.

Less clear-cut is the question of menu-driven systems versus programming languages. For the novice or occasional user, menu-driven systems are clearly preferable as the user is guided through the options at any given stage. This can become tedious for a user who either wants the same analyses to be repeated frequently, or requires more flexibility. Menu-driven systems tend to be inflexible and generally offer fewer facilities because they require extensive hierarchical menus which are slow to operate. Programming languages, on the other hand, permit considerable flexibility but have their own syntax which makes them difficult for the novice or occasional user. This may be offset by the availability of programmers to write user-friendly programs or procedures, particularly in large institutions.

3.3 **On-site testing of packages**

If it is possible to test the package before purchase it is advisable to collect some 'typical' data sets and users. The data files used for demonstration purposes will almost certainly show off the facilities of the package superbly, but they will not reveal its deficiencies. If the package is intended for undergraduates have it tested by them before buying it. If possible carry out the same analyses using an established package that is known to give the correct answers. This will enable the accuracy and integrity of the package to be tested immediately.

4. DATA HANDLING

Data management and manipulation is an important aspect of data analysis. It is also one that can occupy an excessive amount of time and effort.

4.1 **Data entry and storage**

Entering and storing data cannot be wholly separated. Most packages can accept data from ASCII (text) files entered using a system editor or word processor as well as directly from the keyboard. Such files provide convenient data storage and permit operating system editors and other, for example sorting, facilities to be used independently of the package. However, several difficulties can be encountered. Simplest of these is how to denote missing values: not all programs can accommodate missing values, and different packages may require different symbols.

Many packages and programs possess their own storage facilities. Generally they require large amounts of storage space, a lack of which will result in failure of the program. For the large statistical packages such storage facilities are essentially crude databases, although for SAS a database module can be purchased offering highly sophisticated facilities, and SPSS-PC has a data-entry module which can be used separately.

In general, databases such as dBASE, and spreadsheets such as LOTUS 1-2-3 and SUPERCALC provide powerful data management systems which can, if not linked directly to packages, provide ASCII files for subsequent input. They can, in some cases, be persuaded to yield quite sophisticated analyses on their own.

4.2 General data manipulation

It is often necessary or desirable to re-structure the data, for instance sorting, concatenating, splitting columns of data, adding or deleting individual samples and selecting subsets of data. Such operations can be accomplished outside the package or program, but this does entail using a different system and having to store, at least temporarily, extra data. Large packages usually have extensive data manipulation functions, but some smaller packages may offer only limited functions.

4.3 Calculations and transformations

Simple arithmetic calculations are frequently useful during analyses for a variety of reasons. If they are not available or are too restricted, they can be performed outside the package, for instance using a spreadsheet.

One of the primary uses of simple calculations is for transforming data prior to analysis. There are many transformations, and the reasons for using them depend on the data or analysis. They will be covered in later sections, but it is worth noting that specific transformations are often provided.

4.4 Data output

Programs that run in batch on mainframe and minicomputers automatically produce output files which can later be perused or printed. Some packages when run interactively, such as MINITAB and GENSTAT, are able to copy results to an output file at the same time as they are presenting them on the screen. If a program has neither facility it is essential that hard-copy can be obtained, or that it presents a single screen-full of information at a time.

Output files, after editing, provide a means of transferring results from one system to another. Some packages permit greater control over where output is directed. Non-commercial programs can often be easily adapted to provide suitable output files if they are frequently needed, although such freedom cannot be assumed with commercial software.

4.5 Graphics

Graphical output on mainframe and minicomputers is of two distinct types. Most basic are 'line-printer' graphs which are produced as part of a normal ASCII output file and can be printed alongside other results. These are now a standard feature of large packages which provide extensive scaling and labelling facilities. Recent years have also seen the incorporation of device-independent high-quality (plotter-type) graphics. They provide considerable flexibility for plotting graphs, histograms and contour plots on a wide range of devices.

High-quality or high-resolution screen graphics has not been a feature of the large computers because of the high cost of display devices, although it will be in future. Microcomputers have long provided high-resolution screen graphics, and the major microcomputer packages provide output for a range of devices.

Particular graphical points to note are the ability to scale and label axes, and the ability to label points is useful. Examples of line printer and high quality graphics are presented in the following sections which examine how particular common analyses are accomplished using computer packages.

5. ANALYSIS OF VARIANCE

The analysis of variance of designed experiments is one of the most common forms of statistical analyses for testing differences between groups of objects or samples. It is extensively documented (8–11) and these texts provide both a good introduction and a source of continuing reference. However, a necessarily brief description will serve to demonstrate the basic requirements for a good program.

A typical situation might be as follows. An experimenter wishes to test the effects of temperature and salt (the treatments) on the growth rate of bacteria (the response). The experimenter selects the levels of the treatments, in this case 10, 20 and 30°C for temperature and 2 and 4% for salt, and an equal number of observations are taken for each combination of treatments. This is an example of a balanced, orthogonal two-factorial design where all treatments have been equally replicated. Orthogonality and balance are important properties of designed experiments which ensure that the effects of the treatments can be assessed independently. A good package should be able to detect imbalance and non-orthogonality and display warnings if they occur. In this example both balance and orthogonality will be lost if a single observation is lost. In certain circumstances it is permissible to estimate a missing observation, and such routines are often incorporated to allow the balance to be recovered. However, it should not allow more than 50% of the observations to be estimated!

This is a two-factorial or two-way analysis of variance. Most basic packages, such as MINITAB, allow one- and two-way analyses to be performed, but others are more flexible. A typical output is presented in *Table 1* from GENSTAT. This includes an analysis of variance table, tables of means and standard errors. The analysis of variance table partitions the total variability in the data into systematic variation due to the treatments and random variation. The systematic variation is further divided to provide an assessment of the individual effects of salt and temperature and their interaction, for instance synergistic or antagonistic interaction, and the residual entry of the table is the random variation not accounted for by the treatments. The final column of the table headed 'F pr.' presents significance levels for the treatment effects.

Mean values are presented for each treatment, averaged over the levels of the other, and for each combination of treatment levels averaged over the replicates. These should be accompanied by standard errors of either the means or their differences. Only one standard error is presented for each table because the residual variability of observations about the means is assumed to be constant. This is a basic assumption underlying analysis of variance. It can be checked, along with assumptions of normality and independence for formal significance tests of the treatment effects, by plotting the residuals against fitted values. The residuals are calculated as the differences between observed values and their corresponding means in the salt–temperature means table. They should be randomly scattered in an even band about a mean of zero. The band should not, for instance, increase or decrease as the means increase, as shown in *Figure 1*. If it does then none of the assumptions hold, and the standard errors and significance levels are incorrect. The presence of such significance levels does not verify the assumptions—they should be checked by plotting the residuals, or some other test (8).

Table 1. Two-factorial analysis of variance produced by GENSTAT with treatments salt and temperature.

Analysis of variance

Variate: rate

Source of variation	*d.f.*[a]	*s.s.*[b]	*m.s.*[c]	*v.r.*[d]	*F pr.*[e]
Salt	1	55.944	55.944	804.47	<0.001
Temperature	2	114.651	57.325	824.33	<0.001
Salt – temperature	2	0.699	0.349	5.03	0.052
Residual	6	0.417	0.069		
Total	11	171.711			

Tables of means

Variate: rate

Grand mean 10.506

Salt	*2%*	*4%*
	8.347	12.665

Temperature (°C)	*10*	*20*	*30*
	6.993	10.010	14.515

	Temperature (°C)		
Salt	*10*	*20*	*30*
2%	5.020	8.005	12.015
4%	8.965	12.015	17.015

Standard errors of differences of means

Table	*Salt*	*Temperature*	*Salt – temperature*
rep.[f]	6	4	2
s.e.d.[g]	0.1523	0.1865	0.2637

[a]Degrees of freedom.
[b]Sums of squares.
[c]Mean (sums of) squares.
[d]Variance ratio.
[e]*F* probability (significance level).
[f]Replication.
[g]Standard error of difference.

If the residual assumptions do not hold, for example if the response is a probability, it is necessary to transform the response variable in some way. There are various transformations which induce normality or constant variance to ensure validity of the assumptions. The most common of these are logarithmic, square root and reciprocal, but others can be applied depending on the problem (8). The analysis of variance table, residual plots, tables of means and standard errors, and transformations are basic essential features of analysis of variance, regardless of the complexity of the problem. This is a very simple example. In practice, analysis of variance is much more complex and can account for many more treatments, plus various sources of extraneous or background

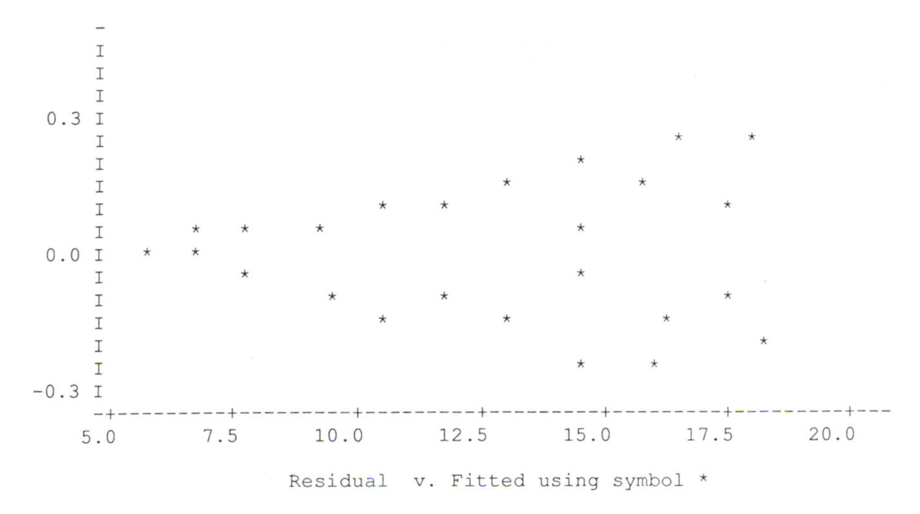

Figure 1. Example of GENSTAT line printer output. A plot of residuals (vertical axis) versus fitted values (horizontal axis) after an analysis of variance. The residual variance is not homogeneous, but proportional to the fitted values.

variability, blocking and co-variates analyses. There are also many designs whose object is to maximize the information gained from the experiment, but minimize the number of observations. It is in such cases that a good analysis of variance procedure or package is invaluable.

6. REGRESSION ANALYSIS AND MODEL-FITTING

This section continues the theme of assessing the effect of several explanatory variables on a dependent variable. It is generally assumed that a mathematical relationship exists between the dependent and explanatory variables, although it must be emphasized that the existence of such a relationship does not imply the existence of a physical or causal relationship. Models are fitted for optimization, description or prediction, and consequently there are many models that can be fitted and various ways of accomplishing it.

These techniques can be put in context by considering several examples. Bacterial growth over time typically produces a sigmoid-shaped curve with log count and time as the dependent and explanatory variables, as in *Figure 2*. The simplest approach to modelling these data is to fit an appropriate polynomial as an approximation to a true underlying, but possibly unknown, function or relationship. An alternative is to derive from theoretical considerations a mathematical form for the relationship. Both will be demonstrated.

6.1 **Linear regression**

The best established and most familiar technique is linear regression. Most introductory statistics books include a section on linear regression, but a more thorough approach is given by Draper and Smith (12); also Caulcutt and Boddy (13) provide an excellent discussion of linear regression in the context of calibration and testing bias in exper-

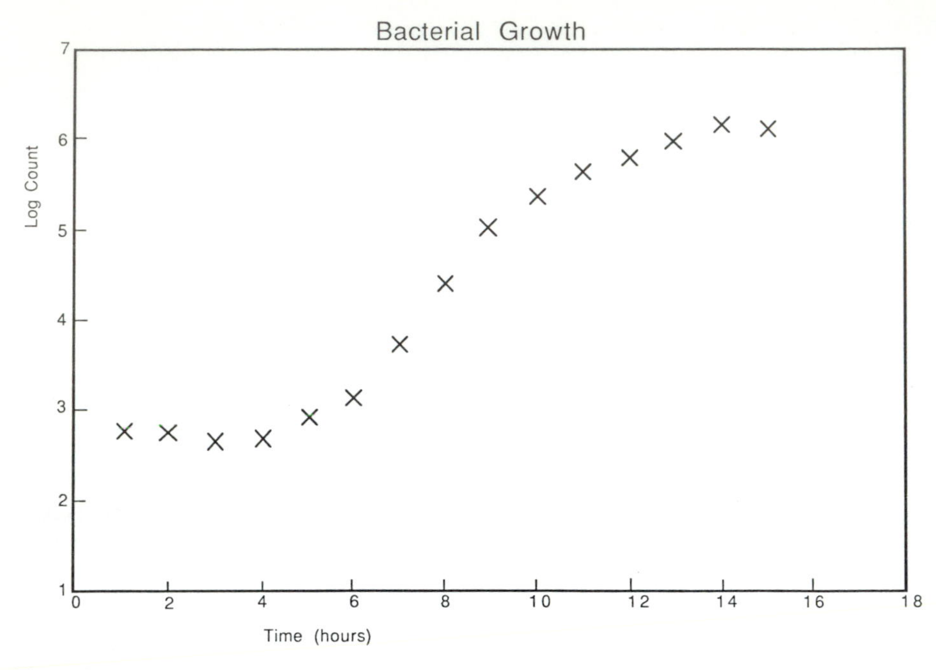

Figure 2. High-quality plot of a typical bacterial growth sequence.

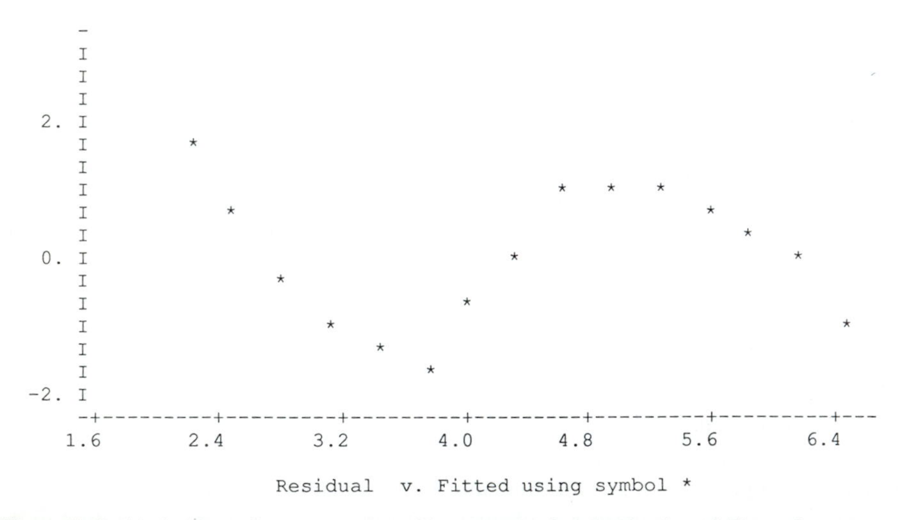

Figure 3. Residuals after a linear regression of log count on time for the data of *Figure 2*.

imental observations. Linear regression is used for fitting linear models such as polynomial curves involving one or more explanatory variables.

The shape of the curve in *Figure 2* suggests that a cubic polynomial could be used. This is a linear model and so can be fitted very easily, and many computer packages incorporate the necessary routines. It can be accomplished by either selecting the appropriate model in some packages, or by adding linear, quadratic and cubic terms

Table 2. Summary analysis produced by GENSTAT for fitting a cubic polynomial curve to bacterial growth in *Figure 2*.

Regression analysis

Response variate: LogCount
Fitted terms: Constant, Time, Time2, Time3

Summary of analysis

	d.f.[a]	*s.s.*[b]	*m.s.*[c]
Regression	3	27.851	9.283
Residual	11	0.250	0.022
Total	14	28.103	2.007
Change	−1	−1.612	1.612

Percentage variance accounted for 98.9

*MESSAGE: the following units have high leverage:

1	0.67
15	0.67

Estimates of regression coefficients

	Estimate	*s.e.*[d]	*t*[e]
Constant	3.325	0.205	16.23
Time	−0.581	0.107	− 5.42
Time2	0.1302	0.0153	8.49
Time3	−0.005305	0.000631	− 8.41

Accumulated analysis of variance

Change	*d.f.*[a]	*s.s.*[b]	*m.s.*[c]	*v.r.*[f]
+ Time	1	26.206	26.206	1148.83
+ Time2	1	0.034	0.034	1.48
+ Time3	1	1.612	1.612	70.68
Residual	11	0.251	0.023	
Total	14	28.103	2.007	

[a]Degrees of freedom.
[b]Sums of squares.
[c]Mean (sums of) squares.
[d]Standard error.
[e]*t* statistic.
[f]Variance ratio.

sequentially to a model in others such as those in Section 2. The latter is ultimately more useful as it can also accommodate multivariable model-building. In this example the model is fitted sequentially beginning with the linear term and the residuals at this stage have been plotted in *Figure 3*. This shows the power of residual plots since the residuals show a clear systematic pattern, rather than a randomly scattered band, due to the inadequacy of the model. Further terms are needed and can be easily added and assessed. Analysis of variance is a form of linear regression, and the analysis of variance table is used for assessing the fit of the model. However, here the systematic variation is not divided into components, but pooled into a single 'regression' entry. The residual plot provides a check of the assumptions and obvious model inadequacy, but it is possible to calculate and test also the effect of adding new terms to the model. *Table 2* presents the analysis of variance table due to fitting the cubic model, the parameter estimates (note the standard errors) and a summary table giving the change in the fit as each term is added.

These four features provide sufficient information for model-building whether it involves several terms in a single explanatory variable, or several explanatory variables. An example of the latter would be an experiment to test and estimate the effect of various environmental variables on bacterial growth, for instance Einarsson and Eriksson (14), and Gibson *et al.* (15). In either case it is essential that changes to the model can be tested.

6.2 Non-linear regression

Non-linear regression is less extensively documented than linear regression; useful texts are by Ratkowsky (16) and Ross (17,18). Like analysis of variance in the previous section, model-fitting is based on the assumption of normally and independently distributed residuals with constant variance. Violation of these assumptions requires the use of transformations, generalized linear models (19) and likelihood techniques (18).

The cubic polynomial is an empirical descriptive model with no theoretical justification. Jason (20), for example, presents a theoretical derivation of the logistic curve, a standard non-linear growth curve, to model the kinetics of bacterial growth. The logistic curve is sigmoid and provides a preferable alternative to the polynomial. Fitting such a curve to experimental data provides both a test of the model, and a model of the data. The results are given in *Table 3*. Note the presentation of an analysis of variance table to test the fit of the model to the data, and parameter estimates with standard errors.

Recent years have seen growing interest in non-linear modelling for microbiological data, and a few examples are provided by Broughall and Brown (21), Phillips and Griffiths (22), So *et al.* (23), Stannard *et al.* (24) and Ratkowsky *et al.* (25). These models are non-linear models which seek to relate growth kinetics to environmental variables such as temperature, salt, pH, etc. Such modelling is more complex than linear modelling because iterative numerical techniques need to be employed which are more difficult to both program and use. In addition to summary analysis of variance tables it is essential that convergence monitoring facilities are available. Frequently a sensible solution does not exist, and in other cases convergence does not provide the best or most optimal parameter estimates. These problems are discussed by Ratkowsky (16), and by Ross (18) in the manual for MLP (maximum likelihood program), a package written specifically for non-linear model-fitting. MLP is an extremely powerful package, but unfortunately difficult to master. Similar features exist in the large packages described in Section 2 except MINITAB.

There continues to be interest in modelling the probability of bacterial growth as a function of environment (26,27). Probabilities are not normally distributed, and consequently require transformation. An alternative approach is to use generalized linear models which takes account of the non-normality (19). Such facilities exist in the packages noted above, and an alternative is GLIM (Generalized Linear Interactive Modelling) written specifically for these models.

Table 3. Summary analysis produced by GENSTAT for fitting a logistic growth curve to bacterial growth data in *Figure 2*.

Non-linear regression analysis

Response variate: Log Count
Explanatory: Time
Fitted Curve: A + C/(1 + EXP(− B*(X − M)))

Summary of analysis

	d.f.[a]	*s.s.*[b]	*m.s.*[c]
Regression	3	27.997	9.332
Residual	11	0.105	0.010
Total	14	28.103	2.007

Percentage variance accounted for 99.5

Estimates of parameters

	Estimate	*s.e.*[d]
B	0.7651	0.0638
M	8.049	0.117
C	3.3797	0.0953
A	2.6653	0.0595

[a]Degrees of freedom.
[b]Sums of squares.
[c]Mean (sums of) squares.
[d]Standard error.

Modelling is an area of microbiology that is likely to grow over the next few years. It is, however, not sufficient to consider only linear models, as these ultimately have limited application. The use of non-linear and generalized linear models does, however, present many problems, not least of which is simply obtaining parameter estimates. A good, well-programmed and flexible package is essential.

7. OPTIMIZATION

It is frequently necessary to optimize the response of a system to its operating parameters. One type of system is a food product where it is necessary to minimize bacterial growth with respect to such factors as salt or nitrite, pre-heating, packaging and storage temperature. Many analytical instruments also present optimization problems, for instance to maximize peak separation with respect to solvents and columns in analytical high-performance liquid chromatography. Response surface methods are a well established means of locating an optimum, but recently interest has focused on the simplex method, particularly for instrument optimization.

7.1 **Simplex optimization**

A simplex is a geometric figure defined by a number of points greater than the number of dimensions in which it lies, for example a triangle in two dimensions, or a cube

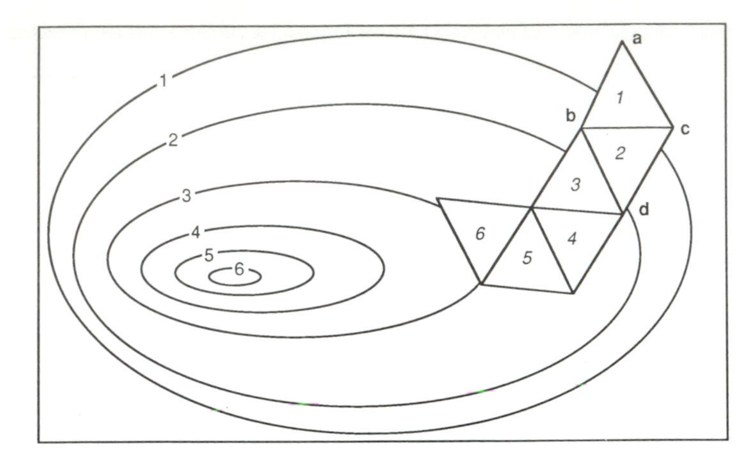

Figure 4. A contour plot of a response surface of a two-parameter system showing the vertices (a, b, c) of the initial simplex, d, the new vertex after the first six iterations of the simplex optimization procedure.

in three. For simplicity the discussion will refer to a two-dimensional system, but the concept extends naturally to higher dimensions. Suppose an instrument's response is to be optimized with respect to two parameters. The performance as the parameters vary can be represented by a contour plot of iso-response lines, as in *Figure 4*. This also shows a number of triangles superimposed on the contours. These present the first six iterations of a simplex optimization procedure.

The first step in the optimization is to define a triangle (numbered 1 in *Figure 4*) in the region and observe the response at the parameter settings defined by the apexes of the triangle (or simplex). These are a, b, c in *Figure 4*. The next simplex is obtained by choosing the least optimal response (a) and reflecting the apex across the opposite face. The response is measured at the new vertex (d), and the process repeated until the optimum has been located. Several rules help to ensure that a sensible result is obtained. If a reflected point has a less desirable response, choose the second least optimal response, and continue as before. An optimal or near-optimal point is identified by a vertex which has been retained for several iterations. The optimality of a solution can be checked by repeating the process from another initial simplex: convergence on the same location gives good indication of a (near) global optimum.

This explanation uses a fixed-size simplex, which is not the most efficient, and several variations, notably modified and super-modified simplexes, provide more rapid convergence (28). These are both variable-sized simplexes whose dimensions depend on the shape of the surface at each iteration.

In *Figure 4* the response surface was assumed to be known. In practice it is not, hence the need for a rapid, efficient method of choosing points at which to take measurements. The computations underlying simplex organization are relatively simple to program and rapid. Several programs exist, notably INSTRUMENT TUNE-UP and CHEOPS, both published by Elsevier Scientific Software, Amsterdam. The limiting factor of simplex optimization is the experimental or data-collecting stage. If it takes a long time to obtain a response at each new vertex, then simplex optimization will be very slow. Instrument tuning is a prime example where simplex optimization is efficient and time-effective.

7.2 **Response surface methods**

If it takes 3 months to obtain a response, which may be the case with the food problem, the best approach to optimization is to obtain sufficient observations simultaneously to estimate the whole surface of the response in the region of interest; or indeed if one wishes to investigate the whole surface, not just the optimum. Response surface methods combine efficient experimental designs, surface estimation and optimum location to address this problem. Again, a two-dimensional system will be used as an example, but the ideas extend naturally to higher dimensions, and extensive details are provided by Khuri and Cornell (29), and Box and Draper (30), as well as in texts on experimental design.

The most popular designs are the central composite or star and cube designs. The pattern of observations form a cube with axial points on each face. Only one observation is taken at each of these points, but replicates are taken at the central point according to certain criteria. The responses form the dependent variable and the parameters are the explanatory variables, and typically the response surface is modelled by a polynomial function, although others can be used. Having fitted the surface it is simply a matter of locating the optimum, although in systems with many parameters or complicated surfaces this may not be entirely straight-forward, and various procedures exist to simplify the problem (29,30).

Response surface models can be fitted by the normal regression techniques discussed above, although some packages possess special response surface options. In fitting the model the main enhancement is partitioning the residual variation into lack-of-fit and pure error. The lack-of-fit is simply the variation of the response (or the mean response for replicated observations) about the fitted surface. Pure error is the variability of replicated observations about their mean: it is experimental error rather than model inadequacy.

Response surface modelling and optimization are considerably enhanced by the use of contour plots and three-dimensional surface plots. Other output is largely the same as for modelling in general, and other comments apply equally well.

8. MULTIVARIATE ANALYSIS

The introduction to this chapter cited several instruments which generate spectra. Such data can be treated as multivariate data in which each variable, whether it is absorbance, peak area or intensity, is the response at a particular wavelength, time or mass. Multivariate techniques permit these variables to be analysed simultaneously and so provide information about their relationships which could otherwise be lost or obscured.

There are essentially three types of technique for reducing dimensionality, recognizing patterns or groups of objects and assessing relationships among and between groups of variables. A good introduction to the subject is given by Chatfield and Collins (31), and Mardia *et al.* (32) and Marriott (33) provide more advanced and detailed treatments. Useful starting points for practical applications to chemical data are the reviews by Brereton (34) and Kowalski (35).

8.1 Matrices and multivariate space

To comprehend multivariate statistics it is necessary to understand the concept of multivariate space. Data can be mathematically represented and manipulated by matrices, but presented graphically to take advantage of human abilities of visual recognition of patterns. It is important to note that good graphics are a prerequisite of multivariate analysis, and interpretation is considerably enhanced if points can be easily labelled. Good three-dimensional and interactive graphics are invaluable in this area. Also a package offering matrix manipulations, but no formal multivariate techniques, can be used for many analyses. But this does require detailed mathematical knowledge of the techniques.

8.2 Multidimensional scaling

Observations on two variables is an unlikely data set, but beyond this it rapidly becomes impossible to represent a full set of data. Multidimensional scaling techniques exploit the structure of data to find ways of presenting them in fewer dimensions without losing or distorting the relationship between points.

Objects measured on three variables lie in a three-dimensional space, but it is possible that their structure is only two-dimensional. The coordinate axes can be rotated so that the points lie on a plane defined by only two of the new axes. In practice it is rare for the points to lie precisely in such a sub-space, but they may lie close to it, that

Table 4. Latent roots and vectors produced by principal components analysis of eight fatty acids. Output produced by GENSTAT.

Principal components analysis

Latent roots

1	*2*	*3*	*4*	*5*	*6*	*7*	*8*
118 659	33 714	7395	5315	254	123	117	66

Percentage variation

1	*2*	*3*	*4*	*5*	*6*	*7*	*8*
71.64	20.35	4.46	3.21	0.15	0.07	0.07	0.04

Trace

165 644

Latent vectors (loadings)

	1	*2*	*3*	*4*	*5*	*6*	*7*	*8*
C12-0	−0.007	−0.002	0.013	−0.008	0.213	0.885	0.385	0.150
C14-0	0.026	0.006	0.136	0.014	0.004	−0.367	0.550	0.736
C16-0	0.274	0.412	0.695	−0.495	0.111	0.009	−0.088	−0.061
C16-1	0.046	−0.062	0.205	0.038	−0.952	0.200	0.049	0.029
C18-0	0.066	0.508	−0.650	−0.517	−0.184	−0.000	0.098	0.051
C18-1	0.522	−0.700	−0.126	−0.468	0.024	−0.011	0.008	0.007
C18-2	−0.801	−0.275	0.116	−0.514	−0.018	−0.004	−0.025	0.036
C18-3	−0.029	−0.019	0.054	−0.012	−0.013	−0.203	0.726	−0.652

is scattered slightly about it, for instance, due to measurement errors. Plotting just the sub-space may not present all of the information in the data, but it may provide a good graphical view of the multidimensional structure without too much distortion.

Principal components analysis (36) and principal coordinates analysis (37) find such sub-spaces. The new axes are the principal axes. The first principal axis passes through the direction of greatest scatter or variance, and so on. They also have the important property of being mutually orthogonal or uncorrelated so that the scatter of points obtained by plotting any pair of principal axes is uncorrelated. The coordinates of the points relative to these new axes are the principal component or coordinate scores.

Table 4 shows the basic results from principal components analysis applied to a set of observations on eight fatty acids. The latent roots are the variances of the principal components; note that they have been placed in descending order of magnitude. The trace is the sum of the latent roots and also equal to the sum of the variances of the original variables. The latent roots have also been presented as percentages of the trace. This provides a guide to the proportion of the total variation in the data explained or accounted for by each component. Also presented are the latent vectors or loadings. These define the rotation of the axes, and graphically they are the cosines of the angles between each principal axis and each original axis.

Graphical presentation of results is a particular feature of principal components analyses. A two-dimensional plot of the data may be obtained by plotting, say, the first two principal components, and the degree of distortion assessed from the latent roots. If the variation accounted for by the first two principal components is greater than about 70% of the total, their plot provides a good representation of the full structure without too much distortion. An example of a principal components plot is given in *Figure 5*. Pyrograms of six strains of bacteria grown on two media were characterized by their

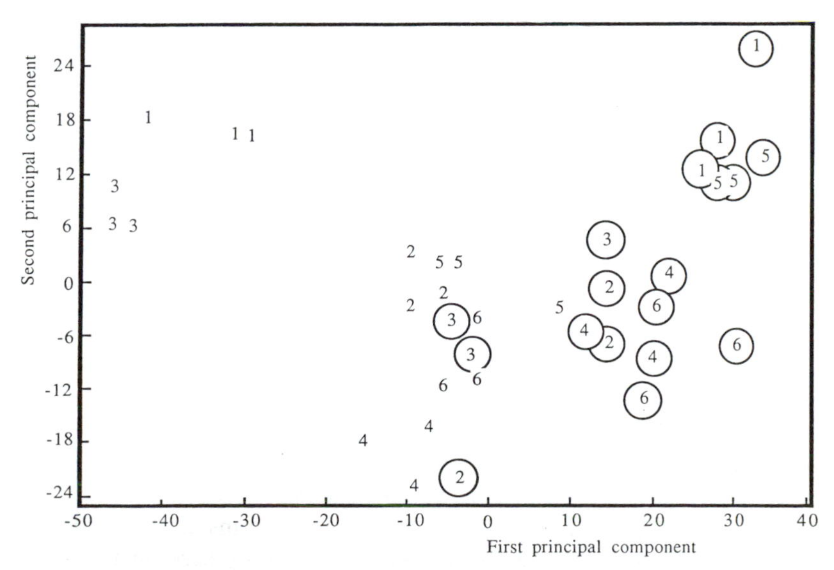

Figure 5. Plot of the first two principal components obtained from pyrograms of six strains of bacteria grown on two media distinguished by circled points. The plot expresses 72% of the total variation. [Reproduced with permission from Gutteridge *et al.* (38).]

peak heights and principal components analysis applied to these variables (38). The principal components plot expressed 72% of the total variation and the tendency of the points to cluster according to medium is immediately apparent and emphasizes the use of such a technique. Plotting the loadings provides information about the relationships between the variables, particularly their correlations. There are several variations on this theme and further details are given by Chatfield and Collins (31), Mardia *et al.* (32), Jolliffe (36) and Gabriel (39).

Principal components analysis is obtained by a latent root and vector (or eigenvalue-eigenvector) decomposition of the co-variance matrix. In the absence of a formal principal components analysis routine, this approach, coupled with matrix multiplication, can be used—as in older versions of MINITAB. Principal coordinates analysis achieves the same results, but it requires a latent roots decomposition of the distance matrix of the objects, and the same comment applies. Both analyses can also be achieved by a singular value decomposition of the original data matrix (36,39).

Principal components and coordinates analyses involve a simple rotation of the axes and are forms of linear multidimensional scaling. Non-linear multidimensional scaling finds axes which may be curved with respect to the orginal axes. There are two basic forms which, like principal coordinates analyses, begin with the distance matrix. The system due to Kruskal (40,41) seeks a low-dimensional representation which attempts to recover the order of magnitude of the inter-sample distances, while that due to Sammon (42) seeks to recover the magnitude of the distances. For an excellent introduction to these methods see Schiffman *et al.* (43).

In multidimensional scaling, the scale of the data can have a strong influence on the results, particularly if one or two variables show extreme variability compared with the others, or if variables are expressed in different units. Autoscaling and logarithmic transformation both scale the data to standardize the variance in all directions and consequently can provide a more balanced analysis. Normalization is a frequently-used transformation to remove sample size effects when the magnitude of results presented by instruments depends on the amount of material introduced. Its effect is equivalent to, for instance, presenting the area of gas chromatographic peaks as a proportion or percentage of their total area. Such transformations should be present as optional features of the analyses and can be accomplished as simple calculations. This applies to all types of multivariate analyses, not just multidimensional scaling.

8.3 **Pattern recognition**

This term is used in its chemometric application and does not imply image analysis. Unsupervised pattern recognition methods such as cluster analysis attempt to group objects. Supervised pattern recognition techniques model known groups of objects for subsequent allocation of unclassified samples.

Cluster analysis, also known as mathematical or numerical taxonomy and widely used in microbiology, is well documented (44–46). Various forms of cluster analysis exist such as hierarchical and optimization techniques and density-based methods. A major difficulty of hierarchical and optimization methods is that the groupings recovered are often influenced by the technique used. It is, therefore, essential that a package offering

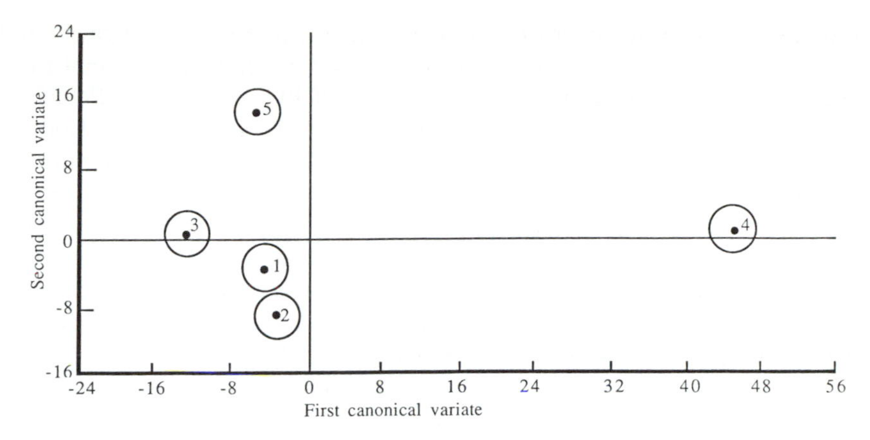

Figure 6. Plot of the genus means relative to the first two canonical variates obtained from pyrograms of five generic groups of bacteria. As both axes are plotted to the same scale the 95% confidence regions are circular. [Reproduced with permission from MacFie *et al.* (49).]

cluster analysis provides several options. For instance, choice of nearest neighbour, furthest neighbour and group average hierarchical methods, and several types of optimization criteria which have a critical effect since the criterion used largely determines the type (shape and size) of clusters recovered regardless of their existence. Density-based methods and fuzzy clustering (47) have the advantage of permitting overlapping clusters and clusters of different shapes and sizes. Fuzzy clustering in particular appears promising but has the disadvantages of being largely inaccessible to non-mathematicians and unsupported by major packages, although it is available in SIRIUS, a pattern recognition package discussed below. The major packages do, however, support a variety of hierarchical and optimization techniques.

The two most popular supervised pattern recognition techniques are canonical variates analysis (31) and SIMCA (48). They are both graphically based with strong links with principal components analysis. Most large packages contain routines for canonical variates analysis which present latent roots, loadings and scores which can be plotted to display the patterns among the various groups. Principal components analysis seeks to maximize distances between all individual objects simultaneously; canonical variates analysis maximizes distances between the group means to provide a clearer basis for subsequent allocations. *Figure 6* shows a plot of the first two canonical variates obtained from 24 peak heights of pyrograms of five generic groups of bacteria (49). Here the five group means have been plotted with 95% confidence regions for each group enabling subsequent sample classification.

SIMCA is more directly related to principal components analysis. Each group or class is modelled independently by a principal components model. The variation in the data is deemed to be composed of systematic and random components, where the large principal components represent the systematic information and the small ones are discarded as random noise. Choice of the correct number of principal components to retain can be made by cross-validation (50). The principal component class model describes the subspace in which the class lies independently of other classes, and allocation of unclassified samples is based on their proximity to this region of space.

SIMCA does not exist as a routine in the statistical packages although it can be programmed around principal components routines. It is a feature of several pattern recognition packages, notably ARTHUR and MICRO-ARTHUR, SIRIUS and UNSCRAMBLER. These contain routines for class modelling, cross-validation and allocation. The last two programs also contain extensive diagnostic routines for detecting objects of high influence which can markedly affect the final results. These packages are written for the IBM PC, computations are generally rapid with operation through a series of menus; but ARTHUR is written for mainframes and operated through a series of commands. They also provide extensive graphical and numerical output.

8.4 **Relationships between groups of variables**

Multiple linear regression (Section 4) is a technique for finding the relationship between a single dependent variable and a group of explanatory variables. Principal components regression (PCR) and partial least squares (PLS) address the same problem (51). In PCR the data are factorized into a number of components representing systematic and random variation, and the random (smaller) components are discarded. The retained components are used in a multiple regression model. This has several advantages over a direct multiple regression. If there are more variables than objects the model cannot be fitted, which is often the case with multivariate calibration of, for instance, near-infrared spectra for determination of fat and moisture. Discarding all principal components with zero latent roots allows a model to be fitted. Discarding the small, random principal components removes some, if not all, of the random variability in the data and provides a more robust model which is less sensitive to the vagaries of sampling. PLS provides a different choice of components to be included in the model and employs an algorithm more suited to microcomputers.

Both techniques are available in SIRIUS and UNSCRAMBLER and incorporate many diagnostic routines for assessing the influence of individual objects. PCR can be carried out in packages which offer both principal components analysis and multiple regression, PLS and cross-validation are much more difficult to program. Both techniques are computationally rapid and can be carried out interactively.

PLS and canonical correlation analysis address the situation where two blocks or groups of variables are to be related. PLS retains the concept of dependency of one set on the other, but both techniques take account of relationships within one group of variables by factorizing them in a similar way to principal components analysis. Both techniques produce sets of scores which can be plotted and sets of loading which define the rotation to new axes. Canonical correlations analysis also produces a set of latent roots which are the correlations between pairs of factors, one from each group of variables. The factors or components are chosen to maximize these correlations so that the technique seeks subspaces in the two blocks which are most highly correlated or most similar.

This form of PLS is incorporated in both SIRIUS and UNSCRAMBLER along with cross-validation and diagnostics. Canonical correlations analysis does not exist in GENSTAT as a standard routine but can be programmed and is available in the standard procedure library supplied with GENSTAT. SAS and STATGRAPHICS both possess routines for this analysis. Programmes are not widely available for applying these techniques to more than two blocks of data, although theoretically it can be done (52).

Procrustes analysis (53) takes a radically different approach. It is based on the assumption that, apart from scale and orientation, the pattern or structure of objects portrayed by the different measured variables are inherently the same. It therefore seeks to scale, rotate and translate one configuration of points to match another; this may be followed by a principal components analysis of the resulting configuration. Generalized Procrustes analysis extends the method to more than two blocks of variables and has proved popular in sensory analysis (54) where the assessors' different understanding and interpretation of sensory attributes lead to apparently different relationships between samples. It can be used in microbiology to find the common information in such disparate groups of measurements as binary tests and gas chromatographic spectra (55,56). GENSTAT and SAS are the only packages to provide a Procrustes analysis routine, although it can be programmed using matrix manipulations.

8.5 **Other techniques**

The previous sections list the more commonly used techniques. Many others exist, the most useful of which are factor analysis, which is similar to, and often confused with, principal components analysis, and multivariate analysis of variance (MANOVA). The former is available in SAS, SPSS and STATGRAPHICS, and the latter is available in SAS, and in GENSTAT as a standard library procedure.

9. CONCLUSIONS

Statistical and data handling techniques are many and diverse. This chapter has attempted to provide the reader with guidelines on the choice of techniques which will be useful and which can be carried out using existing, widely available computer packages and programs. The descriptions and examples have also served to demonstrate the basic abilities and output: frequently techniques are much more flexible and highly developed than has been presented.

A computer package or program should not be chosen on the basis of one class of methods only. It is rare that these will be sufficient. For example, principal components analysis is often a useful adjunct to a series of analyses of variance on several variables as it provides a graphical presentation and summary of the results and a means of resolving apparent contradictions. Similarly, when analysis of variance cannot be applied because a design is unbalanced, multiple regression can be used to recover useful results. Alternatively, multiple regression can be combined with principal components analysis for multivariate calibration, or multidimensional scaling can be used to provide a graphical check of cluster analysis groupings. Other techniques such as exploratory data analysis and non-parametric techniques are also often useful. Lack of flexibility in the statistical package can limit the quality of data collected because the experimental approach and data analysis are not independent—one often implies the other. Therefore, restricted analytical tools can produce inadequate experiments. Restricted analytical techniques can also result in inappropriate and invalid results. For example, linear regression or analysis of variance applied to percentages without prior transformation is not statistically valid and conclusions drawn from such analysis can be seriously misleading. Availability of a basic two-factorial analysis of variance routine often leads to a series of two-factorial experiments where a single four-factor experiment would be more cost- and time-effective and ultimately more informative. It can also result in experiments which do not take account of sources of background variability because the routine cannot perform co-variate analyses.

There is also a more insidious problem to guard against. Berk (2) gives the following comments on documentation: 'Statistical advice is often given or implied, and it is important that it be appropriate. One package makes it easy to compare pairs of means, and the manual suggests that one should pick out the most extreme means and do *t* tests to compare them. There is no suggestion of the need for an *F* test or the adjustment of *p* values (Bonferroni) or the need for an alternative distribution (studentized range)'. The availability of a technique does not necessarily imply that it is either correct or appropriate.

Data analysis requires an understanding of why a particular approach should be used in a given situation and an understanding of the subsequent results. There is currently interest and research into knowledge-based and expert systems which can be linked to packages to provide users with expert advice at each stage of an analysis. A lead in this direction has been taken by GLIM with the GLIMPSE system, and others are likely to follow in the near future (57).

10. REFERENCES

1. Wetherill,G.B. and Curran,J.B. (1985) *The Statistician,* **34**, 391.
2. Berk,K.N. (1987) *The American Statistician,* **41**, 222.
3. Yeo,G. (1984) *The Statistician,* **33**, 181.
4. Lee,J.D. and Lee,T.D. (1982) *Statistics and Numerical Methods in BASIC for Biologists.* Van Nostrand Reinhold.
5. Cooke,D., Craven,A.H. and Clark,G.M. (1981) *Basic Statistical Computing.* Edward Arnold, London.
6. Cohen,L. and Holliday,M. (1982) *Statistics for Social Scientists.* Harper and Row, New York.
7. Thisted,R.A. (1988) *Elements of Statistical Computing.* Chapman and Hall, London.
8. John,J.A. and Quenouille,M.H. (1977) *Experiments: Design and Analysis.* Griffin, London.
9. Cox,D.R. (1958) *Planning of Experiments.* Wiley, New York.
10. Cochran,W.G. and Cox,G.M. (1957) *Experimental Designs.* Wiley, New York.
11. Davies,O.L., ed. (1956) *The Design and Analysis of Industrial Experiments.* Hafner, New York.
12. Draper,N.R. and Smith,H. (1981) *Applied Regression Analysis.* Wiley, New York.
13. Caulcutt,R. and Boddy,R. (1983) *Statistics for Analytical Chemists.* Chapman and Hall, London.
14. Einarsson,H. and Eriksson,S.G. (1986) In *Meat Chilling 1986. Symposium of the International Institute of Refrigeration Commission C2 (Food Science and Technology).* Bristol 10–12 September 1986, IIR, Paris, p. 397.
15. Gibson,A.M., Bratchell,N. and Roberts,T.A. (1988) *Int. J. Food Microbiol.,* **6**, 155.
16. Ratkowsky,D.A. (1983) *Nonlinear Regression Analysis.* Marcel Dekker, New York.
17. Ross,G.J.S. (1975) In *Proceedings of the 40th Session of the International Statistical Institute Warsaw,* Vol. 2, p. 585.
18. Ross,G.J.S. (1987) *MLP Manual.* Numerical Algorithms Group, Oxford.
19. McCullagh,P. and Nelder,J.A. (1983) *Generalized Linear Models.* Chapman and Hall, London.
20. Jason,A.C. (1983) *Antonie van Leeuwenhoek,* **49**, 513.
21. Broughall,J.M. and Brown,C. (1984) *Food Microbiol.,* **1**, 13.
22. Phillips,J.D. and Griffiths,M.W. (1987) *Food Microbiol.,* **4**, 173.
23. So,K., Moneib,N.A. and Kempton,A.G. (1987) *Food Microbiol.,* **4**, 67.
24. Stannard,C.J., Williams,A.P. and Gibbs,P.A. (1985) *Food Microbiol.,* **2**, 115.
25. Ratkowsky,D.A., Olley,J., McMeekin,T.A. and Ball,A. (1982) *J. Bacteriol.,* **149**, 1.
26. Lund,B.M., George,S.M. and Franklin,J.G. (1987) *Appl. Environ. Microbiol.,* **53**, 935.
27. Roberts,T.A., Gibson,A.M. and Robinson,A. (1981) *J. Food Technol.,* **10**, 337.
28. Parker,L.R.,Jr, Cave,M.R. and Barres,R.M. (1985) *Anal. Chim. Acta,* **175**, 231.
29. Khuri,A.I. and Cornell,J.A. (1987) *Response Surfaces.* Marcell Dekker, New York.
30. Box,G.E.P. and Draper,N.R. (1987) *Empirical Model-building and Response Surfaces.* Wiley, New York.
31. Chatfield,C. and Collins,A.J. (1980) *Introduction to Multivariate Analysis.* Chapman and Hall, London.
32. Mardia,K.V. Kent,J.T. and Bibby,J.N. (1979) *Multivariate Analysis.* Academic Press, London.

33. Marriott,F.H.C. (1974) *The Interpretation of Multiple Observations*. Academic Press, London.
34. Brereton,R.G. (1987) *The Analyst,* **112**, 1635.
35. Kowalski,B.R. (1980) *Anal. Chem.,* **52**, 112R.
36. Jolliffe,I.T. (1986) *Principal Component Analysis*. Springer-Verlag, New York.
37. Gower,J.C. (1966) *Biometrika,* **53**, 325.
38. Gutteridge,C.S., MacFie,H.J.H. and Norris,J.R. (1979) *J. Anal. Appl. Pyrolysis,* **1**, 67.
39. Gabriel,K.R. (1971) *Biometrika,* **58**, 488.
40. Kruskal,J.B. (1964a) *Psychometrika,* **29**, 1.
41. Kruskal,J.B. (1964b) *Psychometrika,* **29**, 115.
42. Sammon,J.W. (1969) *IEE Transactions on Computers, C38*, 401.
43. Schiffman,S.S., Reynolds,M.L. and Young,F.W. (1981) *Introduction to Multidimensional Scaling: Theory, Methods and Applications*. Academic Press, New York.
44. Sneath,P.H.A. and Sokal,R.R. (1973) *Numerical Taxonomy*. Freeman, San Franciso.
45. Sneath,P.H.A. (1978) In *Essays in Microbiology*. Norris,J.R. and Richmond,M.H. (eds), Wiley, New York, p. 911.
46. Everitt,B.S. (1980) *Cluster Analysis*. Heinemann, London.
47. Ruspini,E.H. (1973) *Information Sci.,* **6**, 273.
48. Wold,S. (1976) *Pattern Recognition,* **8**, 127.
49. MacFie,H.J.H., Gutteridge,C.S. and Norris,J.R. (1978) *J. Gen. Microbiol.,* **104**, 67.
50. Wold,S. (1978) *Technometrics,* **20**, 397.
51. Geladi,P. and Kowalski,B.R. (1986) *Anal. Chim. Acta,* **185**, 1.
52. Wold,S., Geladi,P., Esbensen,K. and Ohman,J. (1987) *J. Chemometr.,* **1**, 41.
53. Gower,J.C. (1985) In *Handbook of Applicable Mathematics*, Statistics Vol. VIB. Ledermann,W. (eds), Wiley, New York, p. 183.
54. Nute,G.R., Jones,R.C.D., Dransfield,E. and Whelehan,O.P. (1987) *Int. J. Food Sci. Technol.,* **22**, 461.
55. MacFie,H.J.H. (1987) In *Computer Enhanced Analytical Spectroscopy*. Meuzelaar,H.L. and Isenhour,T.L. (eds), Plenum Press, New York, p. 103.
56. O'Donnell,A.G., MacFie,H.J.H. and Norris,J.R. (1988) *J. Gen. Microbiol.,* **134**, 743.
57. Hand,D.J. (1987) *J. R. Stat. Soc., Ser. A,* **150**, 334.

11. APPENDIX

Useful addresses

ARTHUR: Infometrix, Denny Building, 2200 Sixth Avenue, Suite 833, Seattle, WA 98121, USA
Binary: J.B.Evans, Department of Microbiology, University College, Newport Road, Cardiff CF2 1TA, UK
BMDP: BMDP Statistical Software Ltd, Cork Farm Centre, Dannehy's Cross, Cork, Eire, Tel. 021 42722 and BDMP Statistical Software Inc., 1964 Westwood Blvd, Suite 202, Los Angeles, CA 90025, USA, Tel. 213 475 5700
CHEOPS INSTRUMENT TUNE-UP: Elsevier Science Publishers BV, 1 Malenwerf, PO Box 211, 1000 AE Amsterdam, The Netherlands
CRICKET GRAPH: Cricket Software, 30 Valley Stream Parkway, Malvern, PA 19355, USA
Crunch Software Corporation, 5335 College Avenue, Suite 27, Oakland, CA 94618, USA
GENSTAT, GLIM, MLP, NAG routines: Numerical Algorithms Group Ltd, NAG Central Office, 256 Banbury Road, Oxford OX2 7DE, UK, Tel. 0865 511245
MICRO-ARTHUR: R.L.Erksine, British Petroleum, Sunbury-on-Thames, Surrey, UK
MINITAB: CLE.COM Ltd, 97 Vincent Drive, Edgebaston, Birmingham B15 2SQ, UK, Tel. 021 471 4199 and Minitab Inc., 3081 Enterprise Drive, State College, PA 16901, USA, Tel. 814 238 3280
SAS: SAS Software Ltd, Wittington House, Henley Road, Marlow, Buckinghamshire SL7 2EB, UK, Tel. 06284 6933
SIRIUS: O.M.Kvalheim, University of Bergen, N-5000 Bergen, Norway
SPSS: SPSS UK, 9–11 Queens Road, Walton-on-Thames, Surrey KT12 5LU, UK, Tel. 0932 232313, SPSS Europe BV, PO Box 115, 4200 AC Gorinchem, The Netherlands, Tel. 031 1830 36711 and SPSS Inc., 444 Michigan Avenue, Suite 3000, Chicago, IL 60611, USA, Tel. 312 329 3500
STATGRAPHICS: STSC Inc., 2115 East Jefferson Street, Rockville, MD 20852, USA, Tel. 301 984 5123 and Cocking & Drury (Software) Ltd, 155 Friar Street, Reading RG1 1HE, UK, Tel (0734) 588835
STATVIEW: Brain Power Inc., 24009 Ventura Blvd, Suite 250, Caldsas, CA 91302, USA
The National Computing Centre: NCC Micro Systems Centre, 11 New Fetter Lane, London EC4 1QU, UK, Tel. 01 354 4875
The Study Group on Computers in Survey Analysis: 64 Camden Square, London NW1 9XE, UK
UNSCRAMBLER: CAMO A/S, Jarleveien 4, N-7041 Trondheim, Norway, Tel 047 7 51 59 66.

CHAPTER 4

Computers in taxonomy and systematics

SHOSHANA BASCOMB

1. INTRODUCTION

Taxonomy and systematics involve two basic processes.

(i) Classification—the division of a population of individuals, described in terms of a given number of attributes, into a number of clusters.
(ii) Identification—the assignment of an individual, described in terms of all, or some of the above attributes, into one of the previously defined clusters.

Classification and identification encompass both pure and applied research. Thus, classification can be applied to the study of evolution and 'natural' classification as well as to the division of organisms (found in specific environments) into pathogenic, commensal or contaminant taxa.

Various parameters may be used for characterization of bacteria including morphological, biochemical, immunological, molecular, pathogenic and growth requirement aspects of bacterial life, as well as changes arising in the host. Such attributes can be recorded as binary, multistate or continuous data.

A given population may yield different classifications depending on the choice of operational taxonomic units (OTUs), attributes or classification methods. Classification must always precede identification.

Computers have been applied extensively to the study of numerical classification (1) and identification (2); theoretical aspects of the subject have also been reviewed (3,4). The current trend is to apply computers for both data acquisition and analysis, and towards the use of quantitative data. I shall concentrate on the application of computers to data acquisition, preparation of data for analysis, clustering and identification (ID) techniques and the use of computers in some commercial ID systems.

2. DATA ACQUISITION

Numerous attributes are available for taxonomic studies, the inclusion of any of which depends on some criterion of suitability. Thus attributes which show a great degree of variability within the same individual, when tested on different occasions, are not generally useful in taxonomic studies. On the other hand, such attributes may be very important for ID. Thus, resistance to methicillin may be a transient character and not valuable for establishing taxonomic relationships between different staphylococci, even though it is essential for the identification of strains involved in the spread of hospital-acquired infection.

Taxonomic data for populations take the form of tables with attributes (characters, tests, features, variables) occupying the columns whilst OTUs (strains, taxa) form the rows. Time and thought are essential in the design of such taxonomic tables. The number of attributes recorded per strain may be as low as one, or more than 300. Taxonomic information may be entered manually to computer files, or it can be automatically acquired from instruments or other data files. It may be useful to keep the records for an individual strain in lines of less than 80 characters because some data analysis packages restrict the information to this number of alpha-numeric characters per line even though there is no limit to the number of lines. A format as compact as possible is advantageous for data storage on computer but, for eye-ball observation, a spaced format is easier.

2.1 Manual input using editing programs or the extraction of data from dBASE type data banks

Manually recorded attributes must, of course, be entered into data files by hand. It is important to realize that, whenever information is transferred manually, transcription errors occur. An error rate of 1% is probably overly pessimistic and a rate of less than 0.01% is achievable if one is familiar both with the meaning of the data itself and operation of the keyboard. Checking manually entered information is essential, preferably by a second operator.

A number of packages are available for entering information on individual strains to the computer using a pre-designed format. Thus, dBASE III (Ashton-Tate) on an IBM PC/XT, can be used for entering information on culture collection strains (*Figure 1*). The name, designation and width of each attribute (field) is shown as it is entered at the format design stage. Such programs impose limitations on record length and on the total number of records (OTUs). In dBASE III the limits are 128 attributes and 4000 characters per record, and one billion records per file. Other software packages restrict the information to 256 characters per record. If the number of attributes is larger than the package limit, it is possible to have more than one data base file pertaining to the same groups of individuals.

```
Structure for database: a:tax2c.dbf
Number of data records:        7
Date of last update    : 06/16/87
Field  Field Name  Type         Width     Dec
    1  BNUMBER     Numeric          8
    2  SOURCE      Character       12
    3  SOUNUMBER   Character       11
    4  SID         Numeric          6
    5  DATEACQ     Numeric         10
    6  FINALID     Numeric          6
    7  DATETEST1   Numeric         10
    8  TYPETEST1   Character       10
    9  RESULTS1    Numeric         11
   10  TEST1ID     Numeric          6
   11  FITID       Character       10
   12  PROBID      Character        8
   13  DATETEST2   Numeric         10
** Total **                       119
```

Figure 1. Printout of a list of attributes (fields) of a dBASE III file for entering information on culture collection strains obtained by 'DISPLAY STRUCTURE' command.

Alternatively, it is possible to enter data using whatever editor or word processing programs (in non-document mode) are available. Two types of format, fixed or free, may be used. In both cases the sequence of attributes is fixed. In the fixed format each attribute is allocated a definite position in the line. If information for any attribute is missing, the relevant columns are either left blank or filled by symbols indicating missing values. Fixed format is useful for entering test results which are concise and expressed by a fixed number of alpha-numeric characters. Avoid using other keyboard symbols as they may not be acceptable during subsequent processing of the data. If verbal information is to be stored, or if information is patchy with much missing data, a free format method of entering data may be preferred. Here, information on adjacent attributes is entered sequentially, separated either by a comma or a blank space without allocation of spaces for missing information. Both types of file can also be created from a dBASE III file using the 'COPY' commands with either 'SDF' or 'DELIMITED' specification. An example of information in fixed and free formats, obtained from a dBASE III 'dbf' file, is given in *Figure 2*. A microcomputer with a good editor, or dBASE type program is the system of choice for this task. Verbal information, such as species name of the strain, may be given in full but this requires more space and completely accurate spelling. Use numeric or alphabetic codes for genera and species names and other attributes which require verbal description. Artificial intelligence type programming may be needed to cope with patchy verbal information. Missing values should be indicated at the data-entry stage. I have used the maximum possible value for each attribute, for example, 9999 for a 4-digit attribute, as a missing value symbol. String-type characters cannot be used to indicate missing values of a numeric attribute. Some microcomputer editor programs are limited to 500 lines, whilst others are restricted by the total number of characters that can be entered. This may be 32 K or 64 K on some systems. Word processing packages are the least restrictive. Wordstar in non-document mode is an excellent editor with no practical line-memory limitations. Files

```
. copy to a:tax2cc.txt fields bnumber,sid,finalid,results1 sdf
      7 records copied
. type a:tax2cc.txt
   3401      32      23     5215773
   3403     139     139      172444
   3404      23      23     5215773
   3405      41      41      657100
   3427     231     231     5104112
   3428       0      32      174000
   3429       0      32      174000

. copy to a:tax2cd.txt fields bnumber,sid,finalid,results1 delimited
      7 records copied
. type a:tax2cd.txt
3401,32,23,5215773
3403,139,139,172444
3404,23,23,5215773
3405,41,41,657100
3427,231,231,5104112
3428,0,32,174000
3429,0,32,174000

. set printer off
```

Figure 2. Computer output of data files created from a dBASE III 'dbf' file using 'COPY' command ('SDF' or 'DELIMITED') to obtain fixed and free format files.

approaching the maximum number of lines are less easy to manipulate. It is easier to combine two or more files after each has been checked and corrected.

2.2 Automated acquisition of data

Instruments are increasingly in common use to obtain quantitative values for attributes. A simple instrument, used in microbiology, is a photometer which gives a numeric value indicating the amount of light absorbed by a certain sample. Such measurements can be used to indicate enzymatic activity or pH change. Most modern instruments are now provided with facilities for sending such information (directly, or via an interface), to a computer. Information from instruments can be discrete, or continuous.

An example of a discrete type of instrument is the Fluoroskan (Flow Laboratories) used to measure the fluorescence of the 96 wells in multiwell panels. The information sent by the instrument consists of 96 separate numbers sent in the order in which the wells have been read. A continuous reading type of instrument might be a gas–liquid chromatography (GLC) system where the absorbance of the eluting phase is monitored continuously generating a trace which is generally plotted against time.

Interfacing at its simplest level is the ability to connect an instrument that generates data to one that can manipulate or store it. Two problems that arise are the compatibility of the two instruments as regards the type of data transmission—parallel/serial—and the type of connections available. Thus, an IEEE type of connection available on the Commodore PET 4032 cannot be plugged directly into an instrument with the widely used RS232 connections (*Figure 3*). Commercially available interfaces can provide both types of connections and the appropriate electronic circuits. Although 25 possible lines exist in various connectors, only a small number of them are used.

Basic requirements are as follows:

(i) line to send data;
(ii) line to receive data;
(iii) line to indicate readiness to receive data;) Handshaking
(iv) line to indicate sending of signal;) facilities
(v) signal ground pin 7.

It is essential to know which lines perform these tasks.

The following parameters may differ in different instruments and can be catered for by software.

(i) Code used for data conversion, for example Binary, ASCII.
(ii) Number of bits used to convey one piece of information (word length), seven or eight in some computers and 16 or 32 in others.
(iii) Parity (odd, even or no parity).
(iv) Number of stop bits (1 or 2).
(v) Baud rate (300–9600 bits per sec).
(vi) Presence of carriage return and line feed signal—if not, these have to be introduced by software.

In communication packages such as 'CROSSTALK' (Microstuf), or 'DRTALK' (Digital Research), these parameters can be adjusted via one of the screen menus. Interfaces may contain a buffer for temporary storage of data. Many instruments with output ports are not capable of 'handshaking' with the computer. The computer must

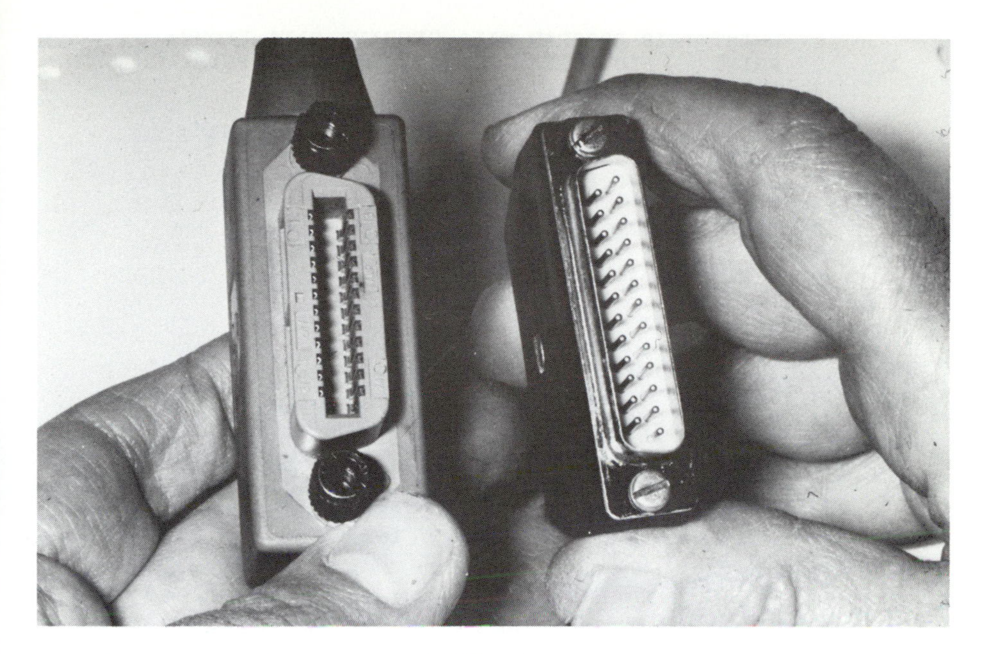

Figure 3. Configuration of RS232 (left) and IEEE (right) connectors.

therefore be ready to accept the information whenever the instrument sends it. This is especially important if the computer is operating in a multitasking mode.

As an example of a system for data acquisition from a discrete instrument, the Fluoroskan can be connected by General Purpose Interface (Small Systems Engineering) to the Commodore PET 4032 with a printer.

Programs for acquiring and printing data can be quite simple (*Figure 4*). This program opens communication lines, ensures the correct configuration of the interface and prints the data. The first part of the program displays the program description on the computer screen (A), then opens channels to the instrument and checks that the buffer in the interface is empty (B), assigns space for the incoming data (C), opens communications to the printer and requests a heading (D), gets data from the instrument (E) and finally prints out all the data received (F). This program is adequate for reading one tray. However, it is prudent at each stage to check that the required process has occurred, thus preventing loss of valuable and irretrievable information. This type of requirement can increase the length of the program considerably.

Chromatograms produced by high-pressure liquid chromatography (HPLC) or GLC instruments generate continuous data where signals sent are absorbance measurements taken at very short time intervals. Such information needs special programs for finding peaks and assigning them to the relevant attributes. Some filtering techniques like Kalman filtering, or Fourier transformation, may help in separating 'true' peaks from 'system noise'. More complex interfaces, for performing some of the data normalizations (see Section 3.1), are necessary for continuous type instruments and include their own microprocessors. Information on interfacing has been reviewed (5), and a program dealing with data acquisition from gel electrophoresis and with the classification of attributes has been published (6).

```
        1   PRINT"THIS PROGRAM WILL RECEIVE DATA FROM";
A       2   PRINT"   FLUOROSKAN AT 1200 BAUD AND PRINT ON LP"
        3   PRINT""

        4   OPEN1,6:OPEN2,5:PRINT#2,CHR$(255);"HE2GA";
B       5   GET#1,P$
        6   IF ST=2 GOTO 8
        7   GOTO 5
        8   PRINT"BUFFER IS CLEAR":CLOSE 1:CLOSE 2

C       9   DIM A$(616)

       10   I=1:PRINT""
       11   OPEN4,4
       12   PRINT"ENTER HEADING"
D      13   INPUT H$
       14   PRINT#4,CHR$(1)H$
       15   PRINT#4,"":PRINT""
       16   PRINT"PRESS START BUTTON ON FLUOROSKAN"

       19   OPEN1,6:OPEN2,5:PRINT#2,CHR$(255);"HE2GA";
E      20   GET#1,A$(I):IF ST<>2 THEN I=I+1
       21   IF I=617 GOTO 40
       30   GOTO 20

       40   FOR I=1 TO 616
       50   PRINT#4,A$(I);
F      55   PRINT A$(I);
       60   NEXT I
       70   CLOSE 1:CLOSE 2: CLOSE 4
```

The program consists of 5 parts performing the following tasks: (A)-prints on the screen the program's description; (B)-opens lines to the interface to clear the buffer; (C)-opens line to printer and prints the heading given by the operator; (D)-gets a preset number of characters from the interface; (E)- prints these on the screen and on the line printer.

Interface codes,line 4:
CHR$(255)-code for the particular interface;
HE2GA-code of data handling: H-baud rate of 1200; E-parity is even; 2-no.of stop bits is 2; G-get mode for obtaining data; A-code for upper case.

Figure 4. Commodore PET 4032 BASIC program for data acquisition.

3. PREPARATION OF DATA FOR ANALYSIS

Binary data require no further manipulation once the information has been entered. Data from instruments may require some normalization procedure to remove any variation not connected with between-strain variability.

3.1 Conversion of machine signals to meaningful data

3.1.1 *Normalization with respect to instrument and occasion variability*

Data acquired on different occasions may show variability due to 'noise' in the system. Small, and large value standards should be used on each occasion to obtain a system/

instrument calibration factor. This factor is used to normalize daily measurements. If this procedure is adopted, slow changes in calibration and drifts in the values may be overlooked. A complete set of standards should be run and regression coefficients calculated once a week to check the accuracy of the calibration procedure. If the system involves the use of buffers they must be prepared afresh every week and included in the weekly system check.

3.1.2 *Normalization with respect to sample size*

This source of error is the most difficult to remove. If plenty of material is available it is advisable to adjust sample size, to a chosen absorbance or to a chosen protein concentration. In the USA, MacFarland standards are commonly used for this purpose. A 0.5 MacFarland standard is equivalent to about 1.5×10^8 colony-forming units (c.f.u.)/ml. When this method of sample preparation relies on judging opacity by eye it is likely to introduce operator error and it is better to use photometer measurement. This facility is often included in commercially available ID systems; in its absence a laboratory photometer is a good alternative. For example, using square 10 mm path cuvettes (Sarstedt No. 67.743, Sarstedt, Leicester LE4 1WA, UK) and an MSE (Measuring and Scientific Equipment, UK) photometer, an absorbance of 0.15 is approximately equivalent to a 0.5 MacFarland standard. The nominal number of c.f.u./ml in such suspensions is 1.5×10^8 but experience suggests that it can vary from 5×10^6 to 3×10^8 depending on the bacterial species. Thus, even with samples adjusted in this way, a several-fold difference in the number of bacteria tested is possible and further corrections may be required.

Individual measurements may be normalized with respect to the following.

(i) The total information available for each OTU (Equation 1)

$$X_{ij} = X_{ij} / \sum_{j=1}^{m} X_{ij} \tag{1}$$

when X_{ij} = value for OTU_i of attribute j; m = number of attributes.

This method is frequently used with pyrolysis GLC/MS data, where the testing provides information on all attributes on each testing occasion.

(ii) One attribute which is present in all individuals, and is usually relatively large (Equation 2)

$$X_{ij} = X_{ij}/X_{ik} \tag{2}$$

where X_{ik} is the value of OTU_i of attribute k which is present in all members of the population.

The numerical values obtained for different attributes may be in different units and of different magnitudes. Use of actual values may diminish the contribution of attributes having small magnitudes. The following two normalizations can diminish such distortions.

(iii) Normalization with respect to standard deviation of each attribute (Equation 3).

$$X_{ij} = X_{ij}/\text{SD}_j \tag{3}$$

where SD_j is the standard deviation of attribute j for the population studied.

(iv) Standardization to Z scores (Equation 4) changes the results for all attributes so that each attribute will have a mean value of zero and a standard deviation of 1.

$$ZX_{ij} = (X_{ij} - \bar{X}_j)/\mathrm{SD}_j \tag{4}$$

where $\bar{X}_j$ is the mean for attribute j in the population studied.

The previous two transformations are only possible after study of a population (at least 30–50 OTUs for each group) to determine the mean and standard deviation of each attribute.

The logarithmic transformation (Equation 5) causes a change in the pattern of distribution of values per attribute and has been found beneficial in taxonomic studies.

$$X_{ij} = \ln (X_{ij}) \tag{5}$$

When using the log transformation it is important to remember that you cannot take a logarithm of a negative value. If the negative values are meaningless in terms of the test and reflect only the 'noise' of the system, change negative values to the smallest positive value of the attribute (Equation 6).

$$\text{If } (X_{ij} \leq 0)\ X_{ij} = A \tag{6}$$

If X_{ij} was recorded as negative or 0 and the mean of this attribute (X_j) is 50, then A is set to 1 and $X_{ij} = A = 1$.

If the negative values have a different meaning from the positive values (for example, decrease and increase in pH in media for testing decarboxylation of amino acids, indicating utilization of glucose or amino acid, respectively) it is legitimate to express the test by two attributes.

The following steps (Equations 7a,b,c) are then necessary:

$$X_{iz} = X_{ij} * (-1) \tag{7a}$$

$$\text{If } (X_{ij} \leq 0)\ X_{ij} = A \tag{7b}$$

$$\text{If } (X_{iz} \leq 0)\ X_{iz} = A \tag{7c}$$

In this way attribute X_{ij} will represent an increase in the value of the original attribute and X_{iz} will represent the decrease in its value.

It is important to realize that all these transformations may themselves introduce error and it is advisable to try more than one transformation on the same data set.

3.1.3 *Creation of artificial attributes*

Sometimes the magnitudes of different attributes are less informative than the ratios between pairs of attributes. New artificial attributes can be created (Equation 8) to exploit such differences.

$$X_{iz} = X_{ia}/X_{ib} \tag{8}$$

where X_{ia}, X_{ib} are values for OTU_i of attributes a and b; X_{iz} is the new artifical attribute.

Artificial attributes may be used instead of, or in conjunction with, existing attributes. The latter increases the total number of attributes and also the weight of those attributes used in creating artificial attributes.

3.1.4 *Assignment of instrument results to attributes*

In pyrolysis GLC/MS, as well as using the absorbance measurements of macromolecules separated on various gels, values for each attribute appear in a fixed order determined by the physical and chemical components of the system. Thus, in GLC of non-derivatized volatile acids, acetic acid will appear before propionic acid. Once the peak values of the chromatogram have been normalized with respect to system noise, a simple program can assign each instrument value to the appropriate attribute. With multiwell panels the content of each well is determined by the experimenter. A fixed layout of attributes can be handled by a relatively simple program. If the layout is changed during the study a more general program is required.

Testing systems that contain synthetic enzyme substrates may suffer both from pipetting errors and from spontaneous deterioration of the substrates. To decrease the error caused by these factors, a substrate set inoculated with saline should be used as controls with each set of test samples. These values can then be subtracted from the test readings to obtain true values for the enzymatic activity of the organism. Alternatively, the means and standard deviations for each substrate control, tested over a reasonable period, can be calculated and the mean + 2 SD of each substrate subtracted from the reading of the attribute (Equation 9).

$$X_{ij} = X_{ij} - (\bar{C}_j + 2^*\mathrm{SD}_{cj}) \qquad (9)$$

where $\bar{C}_j$ is the mean of the substrate control for attribute j, and SD_{cj} is its SD.

Programs for entering the list of OTUs tested on each occasion are also required. One example of automatically acquired fluorescence results for two panels together with the list of OTUs tested on that occasion is shown in *Figure 5*.

```
16105706X B1
2
T1, 460  120
A,   2516, 2595,  28,  47,  367,  381, 2469, 1946,  27,  30,  416, 8609
B,   2503, 2597, 230, 111, 6718,  229, 2317, 2092,  70,  53, 6674, 5629
C,   1846, 1887,  72,  74,  215,  294, 2254, 2678,  52,  61,  682, 2212
D,     10,   11, 281, 478, 8859,  274,   13,   13, 782, 504, 8179,  293
E,    308,  312,  49, 450, 1119,  478,  490, 3294,  44, 920, 1036,  481
F,   4325, 1946, 412, 666, 1456,   96, 3871, 1908, 393, 708, 8807,  108
G,   1164,   22, 586, 280,  309,  284, 1028,   12, 459, 290, 9999,  293
H,   1454,   27, 212, 342,  795, 1428, 1384,   24, 259, 1199, 9999, 1144
T2, 460  120
A,   2359, 1731,  19,  43,  442, 3951, 2460, 2049,  23, 206,  856,  683
B,   2391, 1983, 193,  95, 6740, 5425, 2330, 1978,  71, 122, 6598,  641
C,   2256, 2757,  67,  68,  690, 2373, 2518, 2513,  49,  62, 1008,  505
D,     16,   25, 803, 483, 8637,  311,   13,   15, 2464, 559, 8146,  335
E,    494, 3347,  45, 1087, 1137,  467,  306,  464, 160, 998, 1080,  658
F,   4523, 1963, 496, 765, 8685,  221, 3787, 1949, 604, 1008, 2617,  161
G,   1140,   13, 473, 291, 9999,  268, 1003,   11, 890, 268, 1848,  318
H,   1379,   24, 398, 1100, 9999, 1139, 1314,   25, 346, 1957, 1341, 1320
4
CON,0,32,0
032B1695B,391,321,0
032B1695F,115,213,0
071B2389B,216,83,0
```

Figure 5. Computer output of a data file containing fluorescence measurements of two panels obtained automatically using Fluoroskan (Flow Laboratories), General Purpose Interface (Small Systems Engineering) and PET 4023 computer, dual floppy disc reader and printer (Commodore) and in-house software. Information on strains and results for three additional tests entered manually.

Part of a Commodore PET 4032 BASIC program for assignment of instrument data to attributes performs the following tasks (see Appendix I):

(i) assignment of instrument readings to attribute;
(ii) normalization of well measurements using product-specific coefficients;
(iii) subtraction of the normalized substrate–control readings from normalized substrate–bacteria readings;
(iv) writing of calculated values to a disc file.

3.1.5 *Output format*

Attribute values of taxonomic studies are used in two different ways, eye-ball observation and computer calculation, requiring different presentation of the data. Eye-ball observation is easier if the data are printed in clearly separated columns with headings indicating the identity of attributes and OTUs. For computer handling of data the most compact layout saves memory.

Examples of output of the same data to the printer for eye-ball observation, and to disc file for further computing, are given in *Figure 6a* and *b*. It is essential that the output to disc file includes the testing occasion, or date, for each OTU tested on that occasion. Before attempting any further manipulation of the data it should be checked for errors and omissions.

3.2 **Combining data files**

Testings acquired on different occasions frequently need to be combined into a single file. This is the stage where use of a mainframe computer ought to be considered. Advantages include the ability to handle large files efficiently and the availability, in computing centres, of a variety of sophisticated software, for example, GENSTAT, SPSS^X and SAS. Programs like 'CROSSTALK' or 'KERMIT' (Public domain) may be used for the automated data transfer. Transfer of information from a microcomputer to a mainframe and vice versa can be via a modem or the Packet Assembler Disassembler (PAD), and may be subject to interference. In my experience it is easier to transfer small files separately and to combine the files on the mainframe after checking the integrity of the transferred files. Alternatively, SPSS/PC+ may be used to combine and sort the files using 'SORT', 'JOIN ADD' or 'JOIN MATCH' commands. 'JOIN ADD' combines files containing information on the same attribute for different OTUs; whilst 'JOIN MATCH' combines files containing information on different attributes for the same OTU. The large file is finally transferred to the mainframe using the 'EXPORT' (from the PC version) and 'IMPORT' commands (of the mainframe SPSS^X).

Sorting by one or more attributes is also available on microcomputers and mainframes. For eye-ball examination of the data, sorting the file by bacterial group is an advantage, however, this may not be advisable for certain specific classification routines (see Section 4.3.2). The file is now ready for classification or ID routines.

4. CLUSTERING TECHNIQUES

Cluster analysis may be undertaken to summarize information on a sample in which

a

SUB	ORG CON	ACGAL	ACGLU	ALA	PROT
CON	1.0	82.8	18.8	194.9	0
032B1695B	.4-	64.8-	8.1-	382.2	391
032B1695F	3.2-	14.5-	1.6-	399.0	115
071B2389B	1.6-	63.9-	8.9-	1351.9	216

SUB	ARAB	ARG	ASP	CITR	PROT
CON	9.5	294.8	427.6	142.2	0
032B1695B	2.0-	14.4-	96.9-	35.8	391
032B1695F	1.2-	64.8	85.4-	142.6	115
071B2389B	45.3	147.2	232.7	103.0	216

SUB	BGAL	BGLU	GLUCR	AGLU	PROT
CON	8.7	34.6	19.6	345.2	0
032B1695B	6.8-	23.4-	5.2-	19.8	391
032B1695F	1.2-	6.0-	2.0-	4.5	115
071B2389B	64.8	4.8	4.4-	62.5	216

b

```
CON        16105706X
      1.0      82.8      18.8     194.9       9.5     294.8
    427.6     142.2       8.7      34.6      19.6     345.2
    323.8     488.7     194.1     241.4     260.5    2710.4
    144.5    3577.5     442.8     579.3     114.8     311.6
    143.9      82.4     108.7     189.6     345.2      53.7
    197.2    1402.5      79.0      94.0      41.0       1.0
      4.0    1946.0      22.0         0        32         0
032B1695B  16105706X
      -.5     -64.9      -8.2     382.2      -2.1     -14.5
    -97.0      35.8      -6.9     -23.5      -5.3      19.8
    358.6      32.0       7.6     653.8      37.3     -17.9
    356.3    -275.4     -33.7    2977.1    3924.4    3727.6
   3332.3    2187.0     776.7      14.4       2.2       9.1
      6.8    -284.0    -523.0    -225.0     424.0       0.0
   2804.0     -38.0     -10.0       391       321         0
032B1695F  16105706X
     -3.3     -14.6      -1.7     399.0      -1.3      64.8
    -85.5     142.6      -1.3      -6.1      -2.1       4.5
    486.7      76.3       9.1     579.1      57.9       9.3
    363.1     -89.6       7.6    2928.1    3924.8    3728.0
   1446.2    2104.7     842.4      28.9      -7.7      96.1
    -11.5    -288.0    -628.0    -408.0     501.0       9.0
   2853.0      17.0      -9.0       115       213         0
071B2389B  16105706X
     -1.7     -64.0      -9.0    1351.9      45.3     147.2
    232.7     103.0      64.7       4.8      -4.5      62.5
    418.8     261.7      -8.4     996.4     373.8     -48.2
    605.8    -288.4     -15.4     470.6     623.7     221.5
    122.7     167.2      85.8      47.3     138.1      50.3
     26.7    -107.0    -411.0    -352.0      -5.0       2.0
    158.0       3.0     -11.0       216        83         0
```

Figure 6. (**a**) Part of enzyme data output for eye-ball examination. Labels for enzyme tests (attributes), strain number and protein values with wide spacing on every line of the table. (**b**) Enzyme output disc file for further computer manipulation. No attribute labels, strain number and test occasion appearing only once.

the investigator has previous knowledge of the existence of a number of clusters. Results of such studies would provide the investigator with a way of predicting the attributes of an individual, if its class membership has been established, or of predicting class membership of different individuals sampled from the same population on the basis of results of a set of attributes. Alternatively, cluster analysis can be used to test a hypothesis on, for example, the homogeneity of the population of the assumed clusters, or to generate a hypothesis on the number of clusters present in the population.

4.1 Choice of OTUs and attributes

Before starting cluster analysis, a selection of individuals and attributes must take place. How big should the population be? How many attributes should be included? Should features be binary, multistate or continuous? The answers to these questions depend, to some extent, on the purpose of the classification.

OTUs should of course be 'randomly' chosen and, generally speaking, the bigger the sample the more reliable the cluster analysis will be. However, cluster analysis procedures normally require a fairly large memory and most software packages available have an upper limit to the number of OTUs they can process. The restrictions depend both on the software package and, if microcomputers are used, on the size of memory available. The upper limit for CLUSTAN 2.1 is 999 OTUs and 200 for the number of continuous, and 400 for binary, attributes. With some routines, the number of OTUs is restricted to 60. Using the CLUSTER routine of SPSS/PC+ with 45 OTUs, about 0.3 K of memory is required for every attribute and 0.5 K for every additional OTU. If the existence of a number of groups is suspected or known, then it is useful to have at least 5–10 OTUs for each group and better representation in the more commonly occurring groups. In such a study the population used for cluster analysis is, strictly speaking, not randomly selected.

Choice of attributes is even more complicated. Assumptions laid down in cluster analysis procedure are that attributes should be independent of one another. This is not easy to establish. Thus, the presence of β-galactosidase is necessary for the formation of acid from lactose, but the formation of acid requires the presence of a number of other enzymes. Therefore, although the ONPG test value may influence the value of acid production from lactose, the relationship is, strictly speaking, undefined.

When one is looking for 'natural' clusters, selection of attributes that are as different from each other as possible, is encouraged. But it is worth remembering that the more divergent the attributes used the less homogeneous the resultant groups become.

4.2 Measure of resemblance

After the initial data transformation steps (Section 3.1), one can proceed with classification. The first step in classification is the calculation of the resemblance between each pair of individuals. Resemblance can be expressed in terms of similarity, dissimilarity or in terms of distance. Generally speaking, with qualitative (binary) data, similarity is used, while for quantitative (continuous) data the distance measure is used. Similarity values are on a definite scale range of 0–100%; distance values are unrestricted and can take any positive value.

4.2.1 *Measure of resemblance based on binary attributes*

A number of coefficients have been formulated for estimating resemblance between OTUs using binary attributes.

The simple matching coefficient, where the similarity is obtained by dividing the sum of common positive and common negative attributes by the sum of all attributes tested, is often used. Generally speaking the validity of inclusion of attributes absent in both individuals has been questioned. Thus, the similarity ratio (Jaccard coefficient) excludes negative matches. Other coefficients exist which accord different weights to positive and negative matches. A list of coefficients and formulations for calculating their values are available in the CLUSTAN manual.

4.2.2 *Measure of resemblance based on continuous attributes*

A number of options are available for calculating the distance between OTUs on the basis of quantitative attributes. Basically, the position of each individual is visualized in a multidimensional space where each attribute is represented by one axis (dimension), usually orthogonal to existing axes. Formulations for calculation of a few basic distance measures between OTUs are given in *Table 1*. Options for calculation of some of these resemblance values are available in all packages which provide clustering algorithms.

4.3 **Methods of cluster formation**

These can be hierarchical, optimization or density-seeking techniques.

4.3.1 *Hierarchical clustering methods*

Cluster formation can be divisive or agglomerative, namely, we can look at the one population and divide it into K clusters, K being no bigger than the number of individuals. Or one can start with n clusters, each containing one individual, and join clusters on the basis of a particular criterion. These processes are repeated hierarchically, possibly by steps of a predetermined range of resemblance measure.

Table 1. List of resemblance measurements between two OTUs suitable for continuous variables.

1. City block (Manhattan distance)—sum of absolute differences of all variables.
$$D(X,Y) = \sum_i |X_i - Y_i|$$
2. Euclidean distance in the multidimensional space.
$$D(X,Y) = \sqrt{\sum_i (X_i - Y_i)^2}$$
3. Squared Euclidean distance in the multidimensional space.
$$D(X,Y) = \sum_i (X_i - Y_i)^2$$
4. Cosine of vectors of variables.
$$S(X,Y) = \frac{\sum_i (X_i Y_i)}{\sqrt{\sum_i (X_i^2) \sum_i (Y_i^2)}}$$
5. Chebychev distance metric—the maximum absolute difference in values for any variables.
$$D(X,Y) = \text{Max } |X_i - Y_i|$$
6. Distances in an absolute power metric.
$$D(X,Y) = \left(\sum_i (X_i - Y_i)^p \right)^{1/r}$$
Appropriate selection of p and r may yield many of the above measures.
7. Pearson product—moment correlation coefficient (correlation coefficient r) when values for each variable of one individual are plotted against those of a second individual and linear regression line calculated.
$$S(X,Y) = \frac{\sum_i (X_i - \bar{X})(Y_i - \bar{Y})}{\sqrt{\sum_i (X_i - \bar{X})^2 \sum_i (Y_i - \bar{Y})^2}}$$

Table 2. List of some methods for hierarchical cluster formation.

1.	*Single linkage* Distance between two clusters is that between their closest points. Cluster pairs with the shortest distance are joined.
2.	*Complete linkage* Distance between two clusters is that between their two furthest points.
3.	*Average linkage* (UPGMA unweighted pair-group method using arithmetic average) Distance between clusters equals the average distance between all pairs of cases.
4.	*Error sum of squares (Ward's method)* The squared Euclidean distance of each member of a cluster from the cluster centroid is calculated. Two clusters that merge are those that lead to the smallest increase in the overall sum of the squared within-cluster distances.
5.	*Centroid method* After each step the centroids of the new clusters are calculated. Clusters are joined on the basis of the distance between their centroids.

The list of the most frequently used criteria for hierarchical clustering is given in *Table 2*. Some of these options are available in all packages offering clustering algorithms. A detailed list is available in the CLUSTAN manual.

All the above methods are termed hierarchical methods and require a large number of calculations and a fair-sized memory.

4.3.2 *Non-hierarchical methods*

Optimization techniques have aimed to deal with large populations. The population is divided into K groups. The value of K being determined either by the investigator or reached by optimizing a criterion related to the within- and between-cluster variability. Both K and the assignation of individuals to clusters may be iterated until an optimal solution is achieved.

The default operation of such routines will assign the first OTU to cluster 1, the second to cluster 2 and so on up to K, then it will assign the $(K + 1)$th individual to cluster 1 and so on. The groups thus formed are tested for within- and between-group variability. OTUs are transferred until the between-group variability considerably exceeds that within groups. If the file has been previously sorted by groups, the above method may leac to unsatisfactory groupings and OTUs that can act as centroids for each group should be specified.

Density-seeking protocols examine the distribution of individuals in the multi-dimensional space and divide the space into densely or sparsely populated zones.

Principal component analysis and principal coordinate analysis are also used for classification purposes. In these procedures the aim is to express the information available on all attributes in terms of fewer dimensions. A large proportion of the variability can be accounted for by the first three principal components.

4.4 **Portraying results of clustering techniques**

A clustering procedure starts with individual OTUs described in terms of attributes and produces clusters with a different number of OTUs in each, and with a different degree of resemblance between the clusters. Different ways of portraying the relationship between OTUs and clusters are available. Some experience is required for interpretation of the results obtained.

A shaded similarity matrix can be produced. The OTUs are rearranged so that closely related individuals are placed near to each other. The degree of resemblance is indicated by shading. The darker the shading the more similar are the OTUs. Presence of clusters can be observed, but clusters are not formally produced. Such matrices are easier to grasp if the number of OTUs is less than 100; larger matrices are less useful.

Graphical representation of the results can take the form of a dendrogram (tree) where each individual is represented by a line. When a number of individuals merge the cluster formed is represented by a single line, these lines comprise the branches of the tree.

While dendrograms produce non-overlapping clusters, principle component output may show overlapping between clusters and thus provide further insight into the structure of the data. The relationship is commonly portrayed by plotting the data using two or more principal components.

Most packages have facilities for dendrogram and principle component analysis output; the shaded similarity matrix is less common in the newer software programs.

When optimization techniques are used the following information is available: cluster membership of each OTU, the distance of the OTU from the cluster's centroid, as well as its *K*-nearest neighbours; *K* being determined by the investigator. A graphical representation of this information is available in the BMDP package [see (7)].

Table 3. List of some statistical packages for classification.

Package	*References and suppliers*
ARTHUR	Duewer,D.L., Koskinen,J.R. and Kowalski,B.R. (1975) *Documentation for ARTHUR, Version 1-8-75*. Chemometrics Society Report (updated 1981). Infometrix Inc., Seattle, USA
BMDP	Dixon,W. and Brown,M.B. (eds) (1979) *BMDP-79 Biomedical Computer Programs P-series*. 2nd Printing, University of California Press, Berkeley, CA, USA
CLUSTAN	Wishart,D. (1982) *CLUSTAN User Manual*. 3rd Edition, Edinburgh University, Edinburgh, UK
GENSTAT	Alvey,N.G., Banfield,D.F., Baxter,R.I., Gower,J.C., Krazanowski,W.J., Lane,P.W., Leech,P.K., Nelder,J.A., Payne,R.W., Phelps,K.M., Rogers,C.E., Ross,G.J.S., Simpson,H.R., Todd,A.D., Wedderburn,R.W.M. and Wilkinson,G.N. (1977) *GENSTAT. A Generical Statistical Program*. The Statistics Department, Rothamsted Experimental Station, Harpenden, UK
MASLOC	Kaufman,L. and Masart,D.L. (1981) *MASLOC Users' Guide*. Vrije University, Brussels
NTSYS-pc	Rohlf,F.J. (1987) *Numerical Taxonomy and Multivariate Analysis System for the IBM PC Microcomputer (and Compatibles)*. Ver. 1.30. Applied Biostatistics Inc., 3 Heritage Lane, Setauket, NY 11733, USA
SAS	Helwig,J.T. and Council,K.A. (eds) (1979) *SAS Users' Guide*. SAS Inst. Inc., Carry, NC, USA
SPSS/PC+	Norusis,M.J. (1986) *SPSS/PC+ for the IBM PC/XT/AT*. SPSS Inc., Chicago, IL, USA
TAXPAK	Sackin,M.J., unpublished

4.5 **Statistical packages available**

A number of statistical packages are available for both classification and identification. Some are available on mainframes, others on microcomputers and some are available on both. A decision regarding which to use depends to a certain extent on the size of memory available on your microcomputer. A list of some packages is given in *Table 3*, but such lists become outdated fairly quickly. BMDP, GENSTAT, SAS and SPSS/PC+ are general statistical packages that can be used for manipulation of the data, as well as classification and identification. ARTHUR, CLUSTAN and MASLOC are dedicated to classification and identification (clustering and discrimination in the chemometrics jargon) and require other software for data manipulation. Detailed description of these packages is available elsewhere in this volume and I shall restrict my comments to the applicability of these packages to taxonomy.

I have used SPSS more than any other package and find it the most 'user-friendly' package, BMDP and GENSTAT seem less easy to use. CLUSTAN appears the most comprehensive, as regards range of options at each stage of operation as well as availability of less usual procedures, but it was some time before I could feel at home using it.

The commonest output of classification studies is the dendrogram or tree. To be of value it should be easily comprehensible and contain the resemblance scale and original label for each OTU, including previous cluster membership, if available. ARTHUR and CLUSTAN [see (7)] produce the most comprehensively labelled dendrograms, but CLUSTAN requires that the labels for each OTU be entered separately from the rest of the data; this is tedious. SPSS/PC+ dendrograms are condensed to one scale regardless of the actual resemblance range; this calls for special care when comparing two dendrograms.

Some packages provide vertical dendrograms, the scale in these dendrograms is distorted as each unit on the vertical scale indicates a fusion between two clusters; if two fusions occurred at the same resemblance level they would appear at different levels in the dendrogram. Scale units are not related to resemblance scale.

4.6 **Effect of choice of clustering techniques on classification obtained**

The effects, of choice of procedure for data transformation using SPSS/PC+, on resultant classification, have been compared (*Figure 7A, B* and *C*). Dendrograms were obtained for 45 OTUs belonging to four groups, using 18 continuous characters. All three dendrograms were obtained using a square Euclidean measure of distance and Ward's method for cluster formation. Without any transformation, 40 of the OTUs were joined at the first step, and analysis was completed after nine stages. The Z transformation, namely, transforming the data of each test to have a mean of 0 and a SD of 1, causes some spread of mergings. Only 15 OTUs join at the first stage, analysis being completed after 15 stages. Using the natural logarithm transformation the dendrogram appears more informative. Only nine OTUs join at the first stage, analysis requiring 19 stages.

The effect of the choice of resemblance measures is shown in *Figure 8A* and *B*. City block measure (the sum of absolute differences for all attributes) and Chebychev (the maximum absolute difference between pairs of OTUs) were both based on 42 continuous attributes and produced squashed and spread dendrograms, respectively. The method

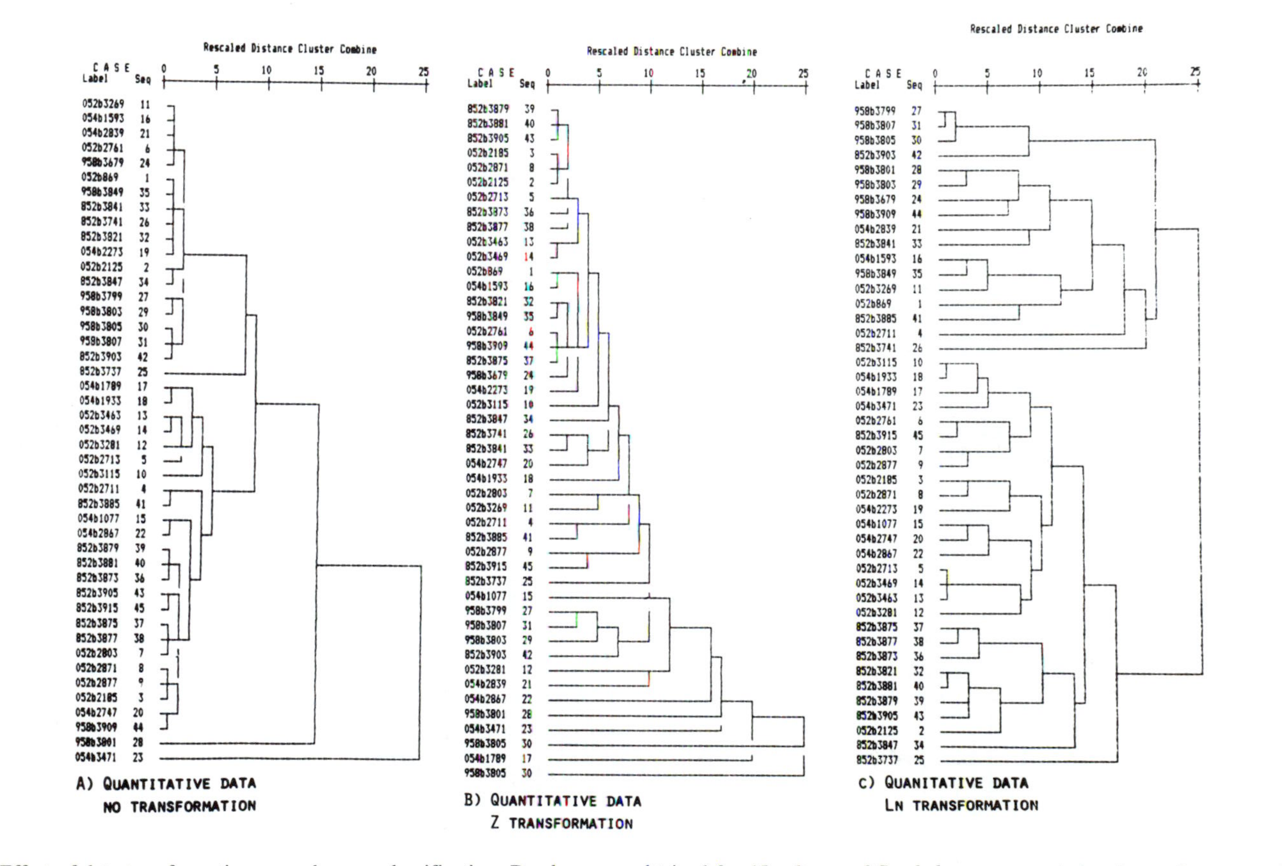

Figure 7. Effect of data transformation procedure on classification. Dendrograms obtained for 45 cultures of *Staphylococcus* spp. belonging to four taxa, all based on the same 18 continuous characters. All distances condensed to an arbitrary 0–25 unit scale. (**A**) No data transformation. (**B**) Z transformation. (**C**) Natural logarithm transformation.

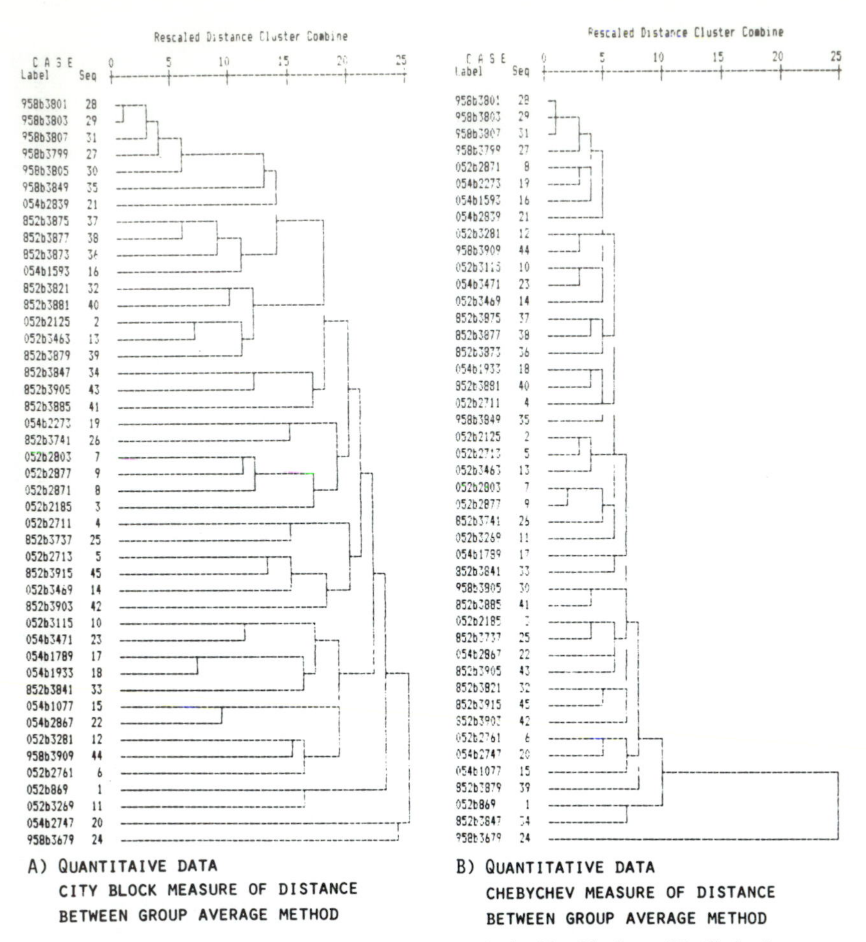

Figure 8. Computer printout showing effect of choice of (**A**) City block or (**B**) Chebychev measure of resemblance on classification.

of cluster formation (*Figure 9*) and type and number of attributes, all affect the resultant classifications.

4.7 **Validation of classifications**

With such a wide range of dendrograms produced from the same set of data one is bound to ask which method should be used. The answer depends, to some extent, on the purpose of the investigation. Generally speaking, more than one option should be examined. Clusters that appear in several classifications are more likely to be valid.

Choice of the final number of clusters in the population and the appropriate level on the resemblance scale are subjective when deciding on a classification. For binary data the similarity value of 85% is often chosen as equivalent to the species cut-off on the similarity scale. Such cut-off levels are not applicable to clustering techniques based on continuous attributes where the distance scales have a variable range. The *acuteness coefficient* introduced by Véron (8) might be applicable here. This coefficient

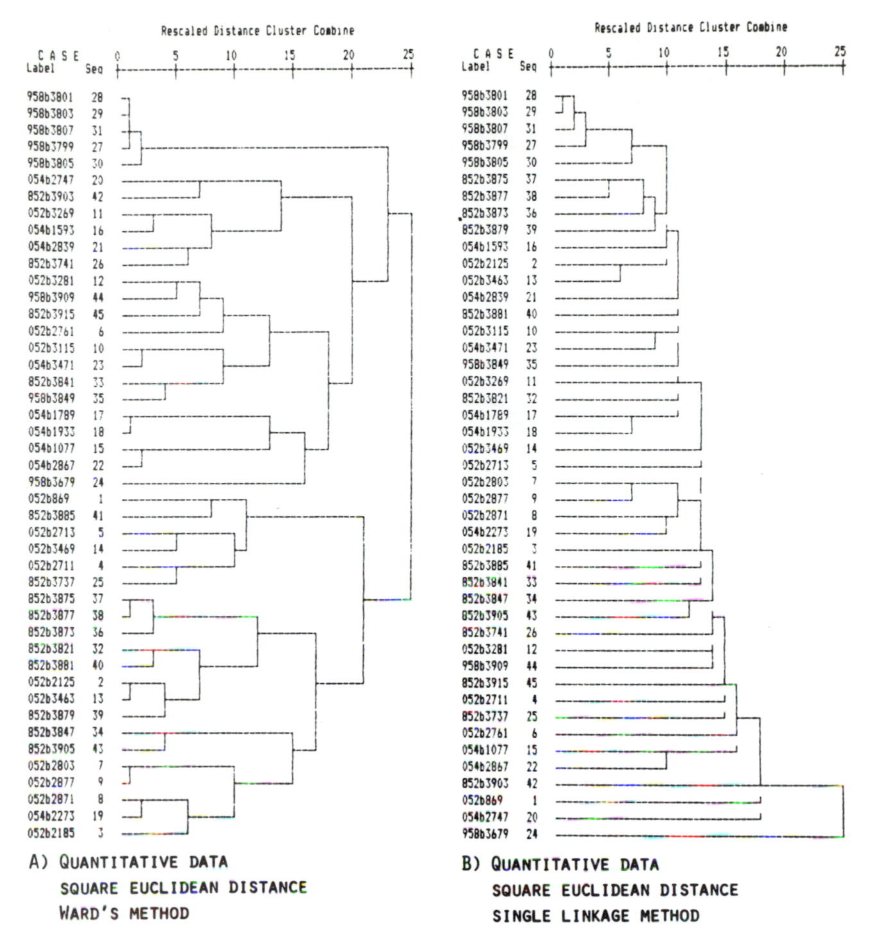

Figure 9. Computer printout showing effect of choice of (**A**) Ward's method and (**B**) single linkage method of cluster formation on classification.

is determined by the number of OTUs in the cluster, the homogeneity of the cluster and the degree of separation from other clusters. Sneath has published a number of programs providing significance tests for the distinctness of the two members of a pair of clusters in Euclidean space (9–11), and one applicable to all dichotomous clusters obtained by the unweighted pair group method with arithmetic averages (12).

A number of tests are used to determine the validity of the formed clusters. The 'RULE' procedure of CLUSTAN can be used to choose the optimal number of non-overlapping clusters. This is achieved by comparing the within- and between-cluster variability. Another way is by charting the distance of individuals from the centroids of each cluster, available in the BMDP *K*-mean clustering routine.

The within- and between-cluster variability can also give an indication of how homogeneous each cluster is and how distant are the clusters from each other. Methods for calculating within- and between-cluster similarities are available in GENSTAT.

Ordination techniques provide a different aspect to the clusters produced, by showing the information on cluster structure, using the value of each OTU on the 1st and 2nd

Table 4. SPSS/PC+ printout of comparison of classification obtained by 'QUICKCLUSTER' compared with identification by API STAPH kit, both using the same binary data for 20-kit tests. Tables produced by 'CROSSTABULATION' command.

Crosstabulation: GROUP By CLUSTMEM cluster from default method

GROUP / CLUSTMEM-> Count	1	2	3	4	Row Total
52		12	1	1	14
54	9				9
852	8		2	4	14
958		4	4		8
Column Total	17 37.8	16 35.6	7 15.6	5 11.1	45 100.0

Crosstabulation: GROUP By CLUSTMEM cluster from default method

GROUP / CLUSTMEM-> Count	1	2	3	4	5	6	Row Total
52		11		2	1		14
54						9	9
852	1		2	1	2	8	14
958				8			8
Column Total	1 2.2	11 24.4	2 4.4	11 24.4	3 6.7	17 37.8	45 100.0

CLUSMEM = clustering obtained by SPSS/PC + 'QUICKCLUSTER'
Group = ID obtained using API STAPH kit

Code for groups: 52 = *Staphylococcus epidermidis*
54 = *Staphylococcus aureus*
852 = *Micrococcus varians*
958 = *Staphylococcus hyicus* ss. *hyicus*

1. 'QUICKCLUSTER' with K (number of groups) = 4
S.epidermidis appearing in clusters 2, 3 and 4
M.varians in clusters 1, 2 and 3
S.hyicus in clusters 2 and 3
Cluster 1 contains all *S.aureus* strains and eight of the *M.varians* strains.

2. 'QUICKCLUSTER' with $K = 6$
Slight improvement in agreement between the two methods of classification, cluster 2 is now homogeneous and all the strains of *S.hyicus* are in cluster 4. However, *S.aureus* and *M.varians* still appear in one cluster.

principal component axes. Circles of the clusters around the centroid (r = averaged within-cluster distance), as well as cluster outlines, can be shown. Andrews's plots (13) can also be constructed when the values obtained for each individual in each test are multiplied by coefficients which can take values between plus and minus π. Another approach will be to compare the classes formed by cluster analysis with cluster membership produced by other means. However, such comparison may show divergent classifications.

A comparison between cluster membership, produced by cluster analysis, and cluster membership produced by identification of staphylococci, on the basis of the same 20 binary tests (API STAPH kit, API France) is shown in *Table 4*, produced using SPSS/PC+ 'QUICKCLUSTER' and 'CROSSTABULATION' procedures.

Identification procedures using the API STAPH kit assigned the 45 individuals into four groups. Using optimization techniques and dividing the population into four or six clusters, did not produce the same class membership of OTUs.

Clustering techniques rarely produce summarized information about the values of the different attributes of each cluster. Bryant (14) described three programs implemented on an IBM PC, designed to evaluate the reproducibility of each attribute in replicate clusters, to summarize the attributes of each cluster and to calculate overlap statistics between major clusters. Alternatively, using the SPSS/PC+ package, the 'FREQUENCIES' command will produce all required statistics for each test. The 'DISCRIMINANT' procedure may be used to validate the clusters formed; choose the most suitable attributes for identification purposes and provide mean and SD for each attribute in each cluster. These procedures will help in choosing the best set of tests.

I have found the SPSS/PC+ the easiest package to use, however, using an IBM/PC/XT with 640 K memory, it could cope with less than 200 OTUs with 80 attributes. Of the packages available on mainframe, the CLUSTAN package appears to offer a large choice of options and procedures.

5. IDENTIFICATION

Once classification has been established it is possible to design an ID system for assigning new (unknown) OTUs to one of the clusters in the matrix. The more frequently ID is performed, the more important it is to make the process as cost effective as possible. ID matrices, which contain fewer attributes and/or require less manual operations, are most convenient for the operator.

ID requires three decisions. Firstly, choice of the taxon which the unknown resembles most. Secondly, calculation of distance between the unknown and the selected taxon compared with distance to other taxa [P(G/X)]; and, thirdly, whether the unknown resembles the chosen taxon sufficiently to be considered a member of this taxon, [P(X/G)].

The effectiveness of an ID system depends not only on its ability to identify unknowns belonging to one of the taxa included in the matrix correctly, but how well it can recognize that an unknown does not belong to any of the taxa in that matrix.

5.1 **ID models for binary data**

A first step in the construction of an ID matrix is the choice of taxa. Many classification studies have been performed with groups defined by 'natural' classification schemes, for example, the enterobacteria, or staphylococci. In contrast, ID schemes are often required to assign newly isolated strains from a certain environment. The composition of taxa found in most environments encompasses more than a single group. Schemes which require one matrix for the identification of all taxa are preferable. Choice of taxa and attributes for an ID matrix may be influenced by the environment from which the unknown OTUs originate. Once the taxa have been chosen, the selection of attributes can proceed.

All ID matrices are constructed so that each pair of groups differs by at least one attribute; however, a single attribute does not allow aberrant behaviour in strains and most ID matrices aim to have a difference of at least two attributes between each pair.

An 'ideal' attribute will be one in which all members of each taxon will give the same result namely, the taxon will be 100% positive or 100% negative. It will also divide the taxa into two equal-sized groups, the first containing taxa positive for that particular attribute, the second containing the negative taxa. If two attributes produce the same partition of the taxa, the information is duplicated and one attribute can be omitted. When attributes are considered collectively, the ability of each attribute to partition the taxa into two groups, which are different in composition from those formed by previously chosen attributes, is the criterion for inclusion in an ID matrix.

If 'ideal' attributes were available, the number required for an ID matrix, with one attribute difference between each pair, would be equal to the square root of the number of groups (15). However, attributes are rarely 'ideal'. They do not give a uniform within-taxon result, only 70–99% of the OTUs give the same result. Also the numbers of positive and negative taxa for each attribute are rarely equal. Many attributes are able to differentiate only a small proportion of the taxa present in the matrix, consequently the number required is often several times greater than the minimal (square root of n) number.

Attribute selection for binary data has been studied (16–19). ID schemes for binary data can be hierarchical, using a dichotomous key, or simultaneous, according equal weight to all attributes. Dichotomous keys still appear in the literature because they are adequate when a small number of attributes are to be considered. Methods for computer construction of such keys are described (19). However, the order in which tests are applied is important, thus, an error (aberrant strain) in an attribute positioned in the first stages of the key will have a greater deleterious effect on final ID than it would had it occurred towards the end. The most logical step was the use of ID tables where all attributes are equally weighted (20).

When large numbers of attributes are used for ID, the pattern of results obtained can be converted to a numerical (octal) code where each digit represents results of three attributes. An index register, where such codes are listed by numerical order, makes ID easier.

Computer ID was originally tried by Dybowski and Franklin (21) but the first successful computer ID scheme reported was by Lapage *et al.* in 1970 (22). ID can be performed by matching the results of the unknown to a library of OTUs and finding the individual OTU which the unknown resembles most (*K*-nearest neighbour), or by matching the unknown to a library of taxa where the summarized results for the taxon are portrayed. In the latter case, the Bayes probability model (22) has been used in most studies and also in commercially available ID systems.

The probabilities of a positive result for each taxon for each attribute are stored in the ID matrix. The likelihood of the unknown belonging to each taxon is calculated by multiplication of the probabilities for all attributes. The likelihood of the unknown belonging to each of the taxa is then normalized with respect to the sum of likelihoods.

The group with the highest normalized likelihood is the one to which the unknown is nearest. The probability that an OTU is nearest to the chosen group is given by this normalized likelihood.

When results of an unknown are compared to those of a number of taxa, it can be assumed, *a priori*, that the unknown will be closer to one taxon than to all others. However, this is not a proof that the unknown is a member of that taxon. Some limits for acceptability of such assignments are required. In the original study (22), two types of limitation were imposed. One relates to a lowest acceptable limit for the normalized score, the other sets an upper limit to the number of attribute disagreements with the chosen group. The first criterion assumes that, if the unknown shows similar distances to more than one group, it is unlikely to belong to any of them. The second criterion puts an upper limit to the deviation from the ideal organism of that group. Conditional probability (21) as well as taxonomic distance, standard error of this distance and pattern distance (23), are examples of more accurate estimates of the resemblance of an OTU to a taxon.

A number of studies suggest that application of artificial intelligence software (24,25) may be able to apply expert knowledge in systems where information on differences between taxa does not lend itself to probability matrix format; for instance, when available for only a limited number of the taxa included in the matrix.

5.2 **ID models for continuous data**

The current trend towards automation and use of analytical instruments in bacteriology yields quantitative data. It is possible to convert these quantitative data to binary or multistate attributes but the conversion inevitably causes loss of information. Methods which are suitable for dealing with quantitative data are required. Although the Bayesian classification rule can be also used for quantitative data, it requires an enormous data bank to provide statistically significant values.

Other methods (26) like the linear discriminant analysis, the *K*-nearest neighbour method, the linear learning machine and SIMCA (27) are more suitable. The partial least squares method (PLS) has been applied by chemometricians to discrimination problems (28).

In all three methods the OTU-by-attribute matrix is visualized in a multidimensional space where each attribute occupies one dimension. In the *K*-nearest neighbour method, an unknown is assigned to the group to which its nearest (in the multidimensional space) neighbour, or neighbours, belongs. This method is the one used when a library of individual cases is kept in the computer memory and the distance of the unknown from each individual case is calculated. The unknown is then assigned to the group that the OTU, from which it is least distant, belongs.

In the linear learning machine the individuals remain in the original multidimensional space, and a line or hyperplane is found by iteration so that it separates the population into two groups in the best way; group one contains a taxon, the second group contains all remaining taxa. An unknown is then assigned according to its position in relation to the hyperplane. The process is almost analogous to the use of dichotomous keys.

In the SIMCA model each group is described by means of its principal component and confidence limits. The principal component is the axis of a cylinder containing 95% of the taxon members, the diameter of the cylinder corresponds to the confidence

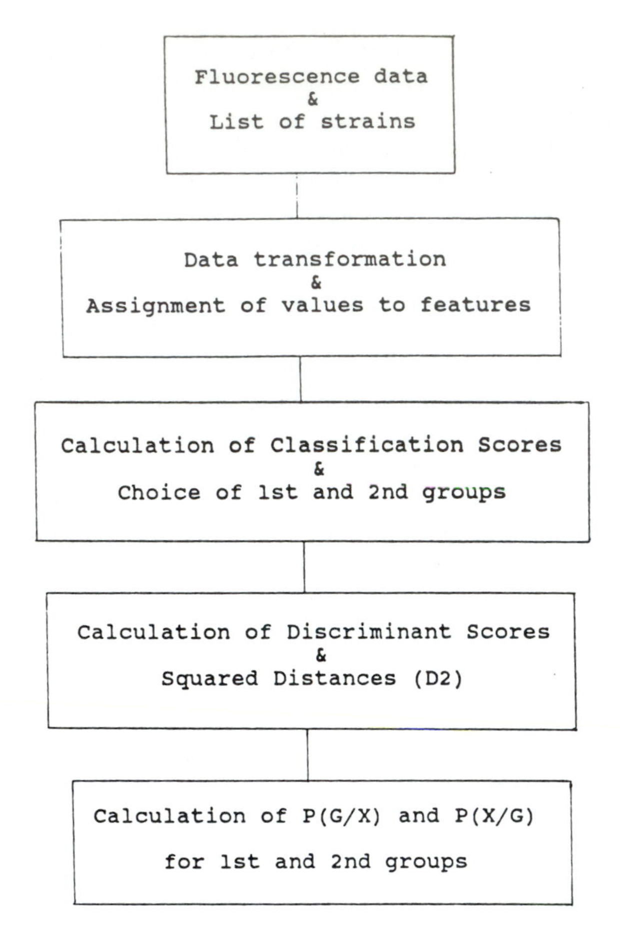

Figure 10. Flow chart of identification procedure using fluorescence measurements and a discriminant function analysis model.

limits of the cluster. In the newer versions capping of the cylinder is also available. SIMCA is the only clustering method that accepts the existence of, and provides categories for, intermediates and outliers.

In linear discriminant analysis the individuals are no longer placed in the original multidimensional space but they are projected onto discriminant lines. The maximum number of lines is dictated by the number of groups. The position of an unknown in the new multidimensional space is calculated and the OTU then assigned to the group to whose centroid it is nearest.

I have tried some of these methods but have concentrated on the discriminant function analysis (DFA) model because I think it is most suitable for use with large numbers of clusters and attributes. This method uses coefficients to multiply each attribute value giving the position of the OTU on a number of discriminant functions. This gives its position in a new multidimensional space, and also its relationship to the position of the centroids of the existing groups.

The main difference between DFA and identification using qualitative data is that

Table 5. Matrices used in discriminant function analysis with their dimensions.

1.	*Classification coefficient* No. of groups * (No. attributes + 1).
2.	*Unstandardized discriminant functions* No. of functions * (No. attributes + 1).
3.	*Group centroids* No. functions * No. groups.

the test values are used more than once in the former and are given different weightings on the different discriminant functions.

The flow chart of ID based on measurements of fluorescence of bacteria–substrate mixtures and the use of DFA is given in *Figure 10*. The experimental data are in a file containing fluorescence data and a list of strains tested on that occasion (as in *Figure 5*). The raw data are transformed and assigned to attributes. Next, the attribute values are used to calculate the classification scores for each taxon and the choice of the most likely group. The third task is calculation of discriminant scores on each discriminant function and calculation of the sum of squared distances from centroids of each taxon on each discriminant function (D^2 and ΣD^2). The last task is estimation of P(X/G), the χ^2 of D^2, and P(G/X), the χ^2 of D^2 divided by the ΣD^2.

The matrices required for ID by DFA are shown in *Table 5*. These matrices can be calculated on an IBM PC/XT using SPSS/PC+ if the number of taxa and attributes are relatively small. Otherwise the calculation can be done on a mainframe computer using $SPSS^X$, BMDP or SAS. The matrices can then be written to a file that can be used on a microcomputer.

An example of assigning two OTUs to one of three possible groups using three attributes and the natural logarithm transformation of attribute values is given in Appendix II.

The first task is choosing the group to which the OTU is nearest. This is achieved by multiplication of transformed attribute values by classification function coefficients of each group to obtain the classification scores for each group. The larger the coefficient the more important is the attribute contribution to the classification score. The group with the largest classification score is chosen.

Thus, in OTU_A the classification score for group 1 is at least twice as large as for the other two groups, and group 1 is chosen as the nearest group, or as the one to which the unknown is most likely to belong. In OTU_B the classification scores for all three groups are actually negative, suggesting that the organism is remote from all three possible groups, but group 2 shows the highest classification score and is therefore the chosen group.

After choosing the nearest group, the fitness of ID is estimated. The DFA proceeds to determine the position of the individual in the discriminant multidimensional space, the probability of the chosen group being the most likely one, P(G/X), and the probability of the unknown belonging to this chosen group, P(X/G). The first probability [P(G/X)] estimates how much the unknown is nearer to the chosen group than to any other group. This probability is similar to the normalized likelihood used with qualitative data. The second probability determines how near is the unknown to the centroid of the chosen

IDENTIFICATION OF STRAINS
TESTED ON 15105706 B1
USING DISK65
UPDATED ON 07/03/86

CON INSUFFICIENT ACTIVITY

GROUP1

STRAIN NUMBER	FIRST GROUP	D↑2	P (X/G)	P (G/X)	SECOND GROUP	P (X/G)	P (G/X)
041B1306B	Ps.aeruginosa	10.5	.651	.999	no growth	.000	.000
023B2084B	Kl.pneumoniae	18.0	.154	.999	Cit.koseri	.000	.000
023B2084F	Kl.pneumoniae	12.4	.489	.999	Cit.koseri	.000	.000
010B3496	E.coli	11.1	.597	1.000	Cit.freundii	.000	.000

GROUP2

STRAIN NUMBER	FIRST GROUP	D↑2	P (X/G)	P (G/X)	SECOND GROUP	P (X/G)	P (G/X)
041B1306F	Ps.aeruginosa	6.7	.658	.999	Sta.epidermis	.004	.000
9990155F	Str.group l	21.2	.011	.475	Sta.epidermis	.008	.315

Figure 11. Printout of identification of several Gram-negative and Gram-positive organisms tested on October 15, 1985. In-house software written in BASIC for Commodore PET 4032 by Mrs Karen J.Foster.

group, or, how much it resembles the ideal organism of that group. These two probabilities are calculated using the Σ D^2 of the unknown from the centroids of all possible groups (in all functions).

This principle was applied to the ID of 87 bacterial taxa of clinical importance using measurements of enzyme activities. The coefficients were calculated on the University of London Computer Centre Amdahl computer using the $SPSS^X$ package; a program for use on the Commodore PET 4032 utilizing these coefficients was written by Mrs Karen Foster. The ID of a number of strains tested on the same day is shown in *Figure 11*, demonstrating the versatility of an enzyme DFA ID system.

Identification is performed in two steps, the unknown is first assigned to one of three groups, basically Gram-positive, Gram-negative 'fermenters' and Gram-negative 'non-fermenters'. In the next step the unknown is identified to the species level.

If total enzymatic activity is below a pre-defined threshold, identification is not attempted and 'Insufficient activity' is printed. If the P(X/G) of the chosen group is less than 0.000001, the unknown is termed as 'Unidentifiable'.

The ARTHUR package is probably the most comprehensive ID package providing different ID models. However, it is also the most 'user-hostile'. ID routines are available in BMDP, CLUSTAN, GENSTAT, SAS and SPSS/PC+.

5.3 Criteria for acceptability of an ID

Criteria for acceptability of an ID relating to either the chosen taxon, P(G/X), or the resemblance to the ideal organism of that taxon, P(X/G), have been suggested. In the original ID paper (22), the limit for P(G/X) based on binary data was greater than or equal to 0.999. In other systems, values ranging from 0.8 to 0.99 have been suggested. With binary data a normalized likelihood of 0.999 is equivalent to, at least, two definitive tests, (0.99/0.01 type) disagreement between the unknown and the chosen group, and between the unknown and the second nearest group. In practice such tests may not be available in the matrix and lower acceptance values may be required. This depends on the number of 0.99/0.01 differences between closely related taxa. It is therefore necessary also to use the second probability. Sneath (23) proposed four different criteria for measurement of similarity of the unknown to the ideal organism. The distance measure (coefficient 2) appears most acceptable as it is not based on any assumption regarding the distribution of OTUs of each cluster around its centroid. The Dybowsky and Franklin (21) conditional probability, where the likelihood of the unknown is divided by the likelihood of the most typical organism, might also be suitable.

With quantitative data, acceptable criteria have only been formulated in the SIMCA model where an unknown is assigned to one group if it is within the 95% confidence limits of only one group. It is termed an intermediate if it is within the confidence limits of more than one group or it is termed as unknown if it is outside the boundaries of all the groups. Outliers are those individuals which are nearer to one group but outside the confidence limits of this group.

Because quantitative results for bacteria quite often show large variability where the SD for attributes within the taxon may approach 100% of the mean, acceptability criteria, used with binary data, may be too strict.

In the DFA model within the SPSS package, an unknown is assigned to a group even if the P(G/X) is only 0.00001 higher than for the second nearest group. No limitation of acceptability on the basis of P(X/G) is imposed. This approach is bound to lead to mis-identification. It might be desirable to choose the value of 0.70 as the lower limit of [P(G/X)] and 0.075 as the lower limit of [P(X/G)] for unequivocal ID.

In cases outside the acceptability levels, the first two likely groups with the relevant P(G/X) and P(X/G) might give an indication whether the organism is an intermediate or outlier, or belongs to a taxon not included in the matrix.

5.4 Additional testing

One of the differences between existing ID models for binary and continuous data is the ability to perform ID using only some of the attributes included in the matrix.

Binary ID systems calculate the probability on the basis of the given data and, if probability levels have not reached the acceptability limit, the attributes which will separate between the most likely groups are selected. Test selection based on ability to separate between the most likely groups, giving two test differences between each possible pair, was described (22). A slightly different approach for test selection has been described (29). Similar models of treatment of missing values are not available for quantitative ID systems. However, as these are often related to instrumental ID systems, where acquisition of data for all attributes is performed simultaneously, the results for all tests are already available at the first attempt at ID.

5.5 Software available for ID

Principles behind the application of Bayes probability model have been described (22,30,31) and have been applied to the identification of different bacterial taxa (14,29,30,32–34). A listing for MATIDN, a general identification program written in BASIC using binary data and a probability matrix, has been published (23). Options for ID based on a nearest neighbor model are available in ARTHUR, CLUSTAN and GENSTAT and could be applied to both binary and continuous data. However, if the

Table 6. Effect of choice of method for calculation of ID coefficients on CPU requirement and ID performance using $SPSS^X$ on an Amdahl computer.

Method	*CPU time*		*Correct ID*
	(min)	*(s)*	%
Direct	0	25	82.4
Wilks	0	26	82.0
Rao	0	27	82.0
Mahal	26	33	82.0

number of groups and attributes is more than 30, this will require an unnecessary waste of resources for repetitive classifications.

Statistical packages like SPSS, BMDP, SAS and ARTHUR contain routines for calculating the discriminant function analysis on the basis of results for OTUs with known group membership, as well as facilities for storing the DFA matrices for further ID.

These packages provide facilities for applying different criteria for inclusion of attributes in the analysis by the use of 'METHOD' commands and selection of a given number of attributes using the 'MAXSTEPS' command. It is also possible to divide the population into a learning set and a test set by the 'SELECT' command. The calculated coefficients can then be used to identify both sets. Output options include the ability to list all cases, giving the first and second most likely groups, the P(G/X) and P(X/G) for both, the discriminant scores and summary tables for agreement between predicted and original IDs for both learning and test sets.

It is important to realize that the different METHOD options require different calculation times. An example of the effect of choice of method on CPU time using $SPSS^X$ on an Amdahl computer is given in *Table 6*.

5.6 Assessment of performance of ID systems

ID systems should be tested with respect to their ability to identify typical, and atypical, members of all taxa included and a few members of taxa not included in the matrix but likely to be mis-identified. This challenge should not include OTUs used in the construction of the matrix. However, this may sometimes be difficult to achieve with taxa for which only a few strains are known.

More importantly, ID systems should be tested with both culture collection and freshly isolated strains as the success rate for the two groups may be different. Reproducibility of ID for a selected number of typical and atypical strains within and between sites should also be established. ID rates for each taxon, and for the whole trial, should be reported as well as the percentage of non-identifiables. Report of agreement should indicate cases where ID was achieved only after using attributes not included in the matrix, for example, serological tests.

ID systems which perform well with all commonly occurring taxa, but are unable to identify rare ones, may not be suitable for a reference laboratory but may still be of use in a routine diagnostic laboratory.

5.7 Commercial ID systems

In all commercial ID systems, the media for performing tests are pre-packaged and all tests are performed simultaneously. The results for tests are also available simultaneously, although in some kits, for example, API STREP, results for part of the set are available after 4 h, the remaining results after overnight incubation.

5.7.1 *ID based on manual reading of test results and Index Code books*

API 20E is an example of such a system. Many of the attributes are similar to the conventional 'biochemical' tests with improved media formulation to provide better reproducibility. Here computers have been used to calculate the ID of the various patterns of results and to print the Index Code books. In these systems ID is determined by the normalized likelihood principle. The reliability of the ID is indicated by the value of the normalized likelihood [P(G/X)], by verbal evaluation, as well as by a number indicating the resemblance of the unknown to the ideal organism. This is expressed differently in the different systems.

5.7.2 *ID based on mechanized recording of test data and on-line ID*

Systems that use multiwell panels as containers for test media, use microcomputers for performing a range of activities. In their simplest form, for example with TouchScan (MicroScan) or Sceptor (Johnston Labs) systems, the results can be entered into a microprocessor memory by touching the correct spot on an illuminated x,y grid having the same dimensions as the panel, or with a moveable x-axis which is applied by the operator to each row in turn. The microprocessor then provides printed ID. Here automation has been applied to the input of data and to ID. The advantage of such a system over the completely manual systems is related to the simplification of the data input operation and the availability of printed results for all attributes. Some of these systems offer facilities for the microprocessor-controlled inoculation of the panels. However, the most important advantage of such instruments is the availability of the microprocessor and the data management package for further processing of the data and for producing epidemiological reports.

5.7.3 *Automated acquisition of data and on-line ID*

A further advance is shown by instruments like Autobac IDX (Organon Technica), AutoMicrobic System (Vitek), AutoScan-4 (MicroScan), Avantage II (Abbott), Cobas-Bact (Hoffman Le Roche) and the AutoScan-W/A (MicroScan) in which reading and recording of test results are both automated. In these systems the media have already been distributed into multiwell panels, or specially designed multichamber cuvettes, which can be used in specific readers so that reading of test results is automated.

Inoculation of the multichamber cartridge is by manual pipetting in the Avantage system, by centrifugation in the Abac (Scalvo, Italy) and Cobas systems, and by application of vacuum and re-pressurization in the AutoMicrobic System. In some the inoculated device has to be placed by hand in the incubator, and then manually transferred to the reading device. In others, manual involvement ends once the cartridge has been placed in the reader incubator. Producing the completely automated instrument is clearly the aim of manufacturers of microbiological instruments. The Automicrobic System was the first such instrument, followed by Cobas-Bact. Recently (December, 1986) MicroScan began selling the AutoScan-Walkaway and API (Analytab Products, USA) showed Aladin.

In all these instruments quantitative measurements are converted to binary data and ID carried out based on the Bayes probability model. The output formats of the different systems vary in clarity and in the amount of information available regarding reliability of the ID.

Two of the instrumental ID systems use the quantitative measurements obtained from the instrument. The Autobac IDX system relies on measurement of growth in the presence of a variety of growth-promoting or growth-inhibiting agents, as well as six qualitative tests determined separately. Identification by a two-stage quadratic discriminant analysis model is performed. The printout consists of values of each attribute and an ID.

The HP 5898A Microbial Identification System (Hewlett Packard) relies on the measurement of derivatized fatty acids using pyrolysis/GLC techniques with a DFA model for ID.

6. CONCLUSIONS

Computers have been introduced into systematics to produce objective classification and to cope with large numbers of OTUs and attributes. Application of computers to ID enables the non-expert to use expertly-designed ID systems. Application of computers to serological or other typing techniques has been limited (35−37). The current trend is towards mechanization and/or the automation of execution of tests, automatic acquisition of data and on-line ID. In spite of availability of mathematical models using continuous data for both classification and ID, binary data have been used predominantly up till now. Hopefully increased availability of quantitative data will be accompanied by a move from the use of simplified inaccurate binary attributes to the more informative quantitative ones.

7. ACKNOWLEDGEMENT

I thank Mrs Karen J.Foster for computer assistance and helpful discussion.

8. REFERENCES

1. Sneath,P.H.A. and Sokal,R.R. (1973) *Numerical Taxonomy*. Freeman, San Francisco.
2. Pankhurst,R.J., ed. (1975) *Biological Identification with Computers*. Academic Press, London.
3. Goodfellow,M., Jones,D. and Priest,F.G. (1985) *Computer-assisted Bacterial Systematics*. Academic Press, London.
4. Goodfellow,M. and Board,R.G., eds (1980) *Microbiological Classification and Identification*. Academic Press, London.
5. Malcome-Lawes,D.J. (1984) *Microcomputers and Laboratory Instrumentation*. Plenum Press, New York.
6. Jackman,P.J.H., Feltham,R.K.A. and Sneath,P.H.A. (1983) *Microbios Lett.*, **23**, 87.
7. Bascomb,S. (1985) In *Computer-assisted Bacterial Systematics*. Goodfellow,M., Jones,D. and Priest,F.G. (eds), Academic Press, London, p. 37.
8. Véron,M. (1974) *Ann. Microbiol. (Inst. Pasteur)*, **125B**, 29.
9. Sneath,P.H.A. (1977) *Classification Soc. Bull.*, **4**, 2.
10. Sneath,P.H.A. (1979) *Computers Geosci.*, **5**, 143.
11. Sneath,P.H.A. (1979) *Computers Geosci.*, **5**, 173.
12. Sneath,P.H.A. (1979) *Computers Geosci.*, **5**, 127.
13. Andrews,D.F. (1972) *Biometrics*, **28**, 125.
14. Bryant,T.N. (1987) *CABIOS*, **3**, 45.
15. Gyllenberg,H.G. (1963) *Ann. Acad. Sci. Fenn. A, IV Biol.*, **69**, 1.
16. Rypka,E.W., Clapper,W.E., Bowen,I.G. and Babb,R. (1967) *J. Gen. Microbiol.*, **46**, 407.
17. Lapage,S.P. and Bascomb,S. (1968) *J. Appl. Bacteriol.*, **31**, 568.
18. Wilcox,W.R. and Lapage,S.P. (1972) *Computer J.*, **15**, 263.
19. Payne,R.W. and Preece,D.A. (1980) *J. R. Stat. Soc. A*, **143**, 253.

20. Cowan,S.T. (1974) *Cowan and Steel's Manual for the Identification of Medical Bacteria*. Cambridge University Press, Cambridge, 2nd edn.
21. Dybowski,W. and Franklin,D.A. (1968) *J. Gen. Microbiol.*, **54**, 215.
22. Lapage,S.P., Bascomb,S., Willcox,W.R. and Curtis,M.A. (1970) In *Automation, Mechanization and Data Handling in Microbiology*. Baillie,A. and Gilbert,R.J. (eds), Academic Press, London, p. 1.
23. Sneath,P.H.A. (1979) *Computers Geosci.*, **5**, 195.
24. Lefebvre,B. and Gavini,F. (1987) In *Abstracts of the 2nd Conference on Taxonomy and Automatic Identification of Bacteria*. Prague, Abstr. No. 56.
25. Valdes,J., Schindler,J. and Matousek,J. (1987) In *Abstracts of the 2nd Conference on Taxonomy and Automatic Identification of Bacteria*. Prague, Abstr. No. 109.
26. Sjøstrøm,M. and Kowalski,B.R. (1979) *Anal. Chim. Acta,* **112**, 11.
27. Wold,S. (1976) *Pattern Recognition,* **8**, 127.
28. Wold,S., Albano,C., Dunn,W.J.,III, Esbensen,K., Hellberg,S., Johansson,E. and Sjøstrøm,M. (1983) In *Food Research and Data Analysis*. Martens,H. and Russwurm,H.,Jr (eds), Applied Science Publications, London, p. 147.
29. Bryant,T.N., Capey,A.G. and Berkeley,R.C.W. (1985) *CABIOS,* **1**, 23.
30. Bascomb,S., Lapage,S.P., Curtis,M.A. and Willcox,W.R. (1973) *J. Gen. Microbiol.*, **77**, 291.
31. Wilcox,W.R., Lapage,S.P., Bascomb,S. and Curtis,M.A. (1973) *J. Gen. Microbiol.*, **77**, 317.
32. Schindler,Z. and Schindler,J. (1983) *Int. J. Biomed. Comput.*, **14**, 17.
33. Williams,S.T., Goodfellow,M., Wellington,E.M., Vickers,J.C., Alderson,G., Sneath,P.H.A., Sackin,M.J. and Mortimer,A.M. (1983) *J. Gen. Microbiol.*, **129**, 1815.
34. Willemse-Collinet,M.F., Tromp,Th.F.J. and Huizinga,T. (1980) *J. Appl. Bacteriol.*, **49**, 385.
35. Thompson,C.J. (1987) *J. Clin. Microbiol.*, **25**, 774.
36. Cinco,M., Dougan,R. and Stefanelli,I. (1977) *Int. J. Syst. Bacteriol.*, **27**, 63.
37. Kramer,S.M., Jewell,N.P. and Cramer,N.E. (1983) *J. Immunol. Methods,* **60**, 243.

9. APPENDIX I

Part of a Commodore PET 4032 BASIC program for assigning, and transformation of instrument data (kept on a disc file) to attributes, and writing the results to a disc file in a format suitable for ID.

```
1 CLR
2 REM FLSKAN PROGRAM
10 PRINT"PROGRAM TO CALCULATE FLUOROSKAN DATA"
20 PRINT"WITH LAYOUT AS FOLLOWS:-"
25 PRINT""
30 PRINT"1)THE MEAN VALUE FOR TRAY CONTROL(2H     AND 8H)
   IS SUBTRACTED FROM ";
35 PRINT"EACH TRAY       SEPARATELY"
36 PRINT""
37 PRINT"2)USE MEAN 4MU AND MCA EQUATIONS TO     CONVERT DATA
   FROM 3A TO 6G";
40 PRINT"AND 9A TO 12G TO PRODUCT CONTROL"
45 PRINT""
50 PRINT"3)SUBTRACTS CONTROL FROM EACH ORGANISM"
55 PRINT""
60 PRINT"4)SAVES CALCULATED DATA IN A DISK FILE"
61 PRINT"":PRINT""
62 PRINT"PUT DISK WITH RAW DATA IN DRIVE 1"
65 PRINT""
69 PRINT""
70 CLOSE 1:CLOSE 2:CLOSE 3
80 DIM T(15,8,12),B$(31,4),U(8,12),SD(2),ME(2)
85 DIM B1(20),C(20,2),A(30,56),BS$(31),O1(50),C1(5)
90 DIM XA(4),YC(4),DI(4),R$(10)
91 R$(1)=" ":R$(9)="         "
92 R$(2)="  ":R$(8)="        "
93 R$(3)="   ":R$(7)="       "
94 R$(4)="    ":R$(6)="      "
95 R$(5)="     "
96 R$(10)="          "
```

```
97 DIM W(30)
100 REM READ TRAY INFO FROM DISK
110 INPUT "NAME OF FILE";E$
120 FL$="1:"+E$
130 PRINT"■"
140 OPEN1,8,2,+FL$+",SEQ,READ"
150 IF DS<>62 GOTO 180
160 PRINT"FILE NOT FOUND"
161 CLOSE 1
162 OPEN1,8,15
163 PRINT#1,"I"
164 CLOSE 1
165 PRINT"RE-ENTER NAME OF FILE"
166 INPUT E$
170 GOTO 120
180 INPUT#1,D$
190 PRINT D$
200 INPUT#1,N
210 PRINT N
220 FOR I=1 TO N
230 GET#1,Z$
240 IF Z$=CHR$(13) GOTO 270
250 T1$=T1$+Z$
260 GOTO 230
270 PRINT T1$
280 T1$=""
330 FOR X=1 TO 8
331 GET#1,A$
332 GET#1,A1$
340 INPUT#1,X1,X2,X3,X4,X5,X6,X7,X8,X9,Y1,Y2,Y3
341 T(I,X,1)=X1
342 T(I,X,2)=X2
343 T(I,X,3)=X3
344 T(I,X,4)=X4
346 T(I,X,5)=X5
348 T(I,X,6)=X6
350 T(I,X,7)=X7
352 T(I,X,8)=X8
353 T(I,X,9)=X9
355 T(I,X,10)=Y1
357 T(I,X,11)=Y2
360 T(I,X,12)=Y3
370 PRINT A$;:PRINT A1$;
390 FOR J=1 TO 11
400 PRINT T(I,X,J);
410 PRINT",";
420 NEXT J
430 PRINT T(I,X,12)
440 NEXT X
450 NEXT I
455 REM READ IN ORG,PROT AND TWO MANUALLY ENTERED TESTS
460 INPUT#1,N2
470 IF ST=2 GOTO 472
471 GOTO 480
472 N2=N*2
473 B$(1,1)="STD":B$(2,1)="CON"
474 FOR K=3 TO (N2+1)
475 B$(K,1)="ORG"+STR$(K-2)
476 NEXT K
```

```
477 GOTO 671
480 PRINT N2
490 FOR L=1 TO N2
500 GET#1,Y$
505 PRINT Y$;
510 IF Y$=CHR$(44) GOTO 550
520 IF Y$=CHR$(13) GOTO 670
530 B$(L,1)=B$(L,1)+Y$
540 GOTO 500
550 REM PROTEIN PRESENT
551 O=1:REM*O=0 NO PROTEIN,O=1 PROTEIN DATA AVAILABLE*
560 GET#1,Y$
570 PRINT Y$;
575 IF Y$=CHR$(44) GOTO 610
580 IF Y$=CHR$(13) GOTO 670
590 B$(L,2)=B$(L,2)+Y$
600 GOTO 560
610 REM TEST1 PRESENT
611 GET#1,Y$
612 PRINT Y$;
613 IF Y$=CHR$(44) GOTO 620
614 IF Y$=CHR$(13) GOTO 670
615 B$(L,3)=B$(L,3)+Y$
616 GOTO 611
617 REM TEST2 PRESENT
620 GET#1,Y$
621 PRINT Y$;
630 IF Y$=CHR$(44) GOTO 643
640 IF Y$=CHR$(13) GOTO 643
641 B$(L,4)=B$(L,4)+Y$
642 GOTO 620
643 S=LEN(B$(L,3))
644 J=1
645 L1=1
646 FOR L2=1 TO S
647 G$=MID$(B$(L,3),L2,1)
648 H=ASC(G$)
649 IF H=32 GOTO 653
650 C1(L2)=H-48:J=J+1
651 L1=L1+1
652 GOTO 654
653 S=S-1
654 NEXT L2
655 O1(L)=0:IF L=1 GOTO 660
656 FOR L3=1 TO (S-1)
657 C1(L3)=C1(L3)*(10↑(S-L3))
658 O1(L)=O1(L)+C1(L3)
659 NEXT L3
660 O1(L)=O1(L)+C1(S)
670 NEXT L
671 N2=N2+1
680 REM SUBTRACT MEAN TRAY CON(2H AND 8H)
    FROM REST OF EACH TRAY SEPARATELY
690 FOR L=1 TO N
695 C(L,1)=T(L,8,2)
696 C(L,2)=T(L,8,8)
700 B1(L)=(T(L,8,2)+T(L,8,8))/2
710 FOR I=1 TO 8
715 FOR J=3 TO 6
```

```
720 T(L,I,J)=T(L,I,J)-B1(L)
721 T(L,I,J+6)=T(L,I,J+6)-B1(L)
725 NEXT J
730 NEXT I
735 NEXT L
740 FOR L=1 TO N
745 FOR I=6 TO 8
750 T(L,I,1)=T(L,I,1)-B1(L)
755 T(L,I,7)=T(L,I,7)-B1(L)
756 NEXT I
757 NEXT L
760 REM FOR A32 TO A36 COL2-1,COL8-7
770 FOR L=1 TO N
780 FOR I=1 TO 5
800 T(L.I,2)=T(L,I,2)-T(L,I,1)
810 T(L,I,8)=T(L,I,8)-T(L,I,7)
820 NEXT I
830 NEXT L
840 REM ARRAY U(ROW NO(1-8),COL NO(1-12))=NO OF
    EQUATION USED(1-2),0=NONE
845 M=0
850 U(1,3+M)=1:U(2,3+M)=1
855 U(3,3+M)=1:U(4,3+M)=2
860 U(5,3+M)=1:U(6,3+M)=2
865 U(7,3+M)=2:U(8,3+M)=2
870 U(1,4+M)=1:U(2,4+M)=1
875 U(3,4+M)=1:U(4,4+M)=2
880 U(5,4+M)=2:U(6,4+M)=2
885 U(7,4+M)=2:U(8,4+M)=2
890 U(1,5+M)=2:U(2,5+M)=1
895 U(3,5+M)=2:U(4,5+M)=1
900 U(5,5+M)=1:U(6,5+M)=1
910 U(7,5+M)=1:U(8,5+M)=1
920 U(1,6+M)=1:U(2,6+M)=1
930 U(3,6+M)=1:U(4,6+M)=2
940 U(5,6+M)=2:U(6,6+M)=2
950 U(7,6+M)=2:U(8,6+M)=0
955 IF M=6 GOTO 970
960 M=6:GOTO 850
970 REM 4MU EQ Y=.405X
980 REM MCA <=1300 Y=.763X:Y=60.388+.60935X
990 REM CONVERT USING RELEVANT EQUATION
1000 FOR L=1 TO N
1010 FOR I=1 TO 8
1020 FOR J=3 TO 6
1030 IF U(I,J)=0 GOTO 1110
1040 IF U(I,J)=2 GOTO 1070
1050 T(L,I,J)=T(L,I,J)*.405
1055 T(L,I,J+6)=T(L,I,J+6)*.405
1060 GOTO 1110
1070 IF T(L,I,J)>1300 GOTO 1090
1080 T(L,I,J)=T(L,I,J)*.763
1081 GOTO 1092
1090 T(L,I,J)=(T(L,I,J)*.60935)+60.388
1092 IF T(L,I,J+6)>1300 GOTO 1105
1093 T(L,I,J+6)=T(L,I,J+6)*.763
1094 GOTO 1110
1105 T(L,I,J+6)=(T(L,I,J+6)*.60935)+60.388
1110 NEXT J
```

```
1120 NEXT I
1130 NEXT L
1140 REM SUBTRACT SUB CON FROM ORGANISMS
1150 FOR L=1 TO N
1160 FOR I=1 TO 8
1170 T(L,I,9)=T(L,I,9)-T(1,I,3)
1180 T(L,I,10)=T(L,I,10)-T(1,I,4)
1190 T(L,I,11)=T(L,I,11)-T(1,I,5)
1200 T(L,I,12)=T(L,I,12)-T(1,I,6)
1210 NEXT I
1220 NEXT L
1230 FOR L=2 TO N
1240 FOR I=1 TO 8
1250 T(L,I,3)=T(L,I,3)-T(1,I,3)
1260 T(L,I,4)=T(L,I,4)-T(1,I,4)
1270 T(L,I,5)=T(L,I,5)-T(1,I,5)
1280 T(L,I,6)=T(L,I,6)-T(1,I,6)
1290 NEXT I
1300 NEXT L
1400 REM OUTPUT SECTION
3810 N3=N2-2
3820 REM STORE ON DISK IN CORRECT FORMAT FOR ID PROG
3830 I=1:L=1:K=6:J=1
3840 FOR M=1 TO 8
3850 A(I,J)=T(L,M,3+K)
3860 J=J+1
3870 NEXT M
3880 FOR M=1 TO 8
3890 A(I,J)=T(L,M,4+K)
3900 J=J+1
3910 NEXT M
3920 FOR M=1 TO 8
3930 A(I,J)=T(L,M,5+K)
3940 J=J+1
3950 NEXT M
4000 FOR M=1 TO 8
4010 A(I,J)=T(L,M,6+K)
4020 J=J+1
4030 NEXT M
4040 FOR M=1 TO 7
4050 A(I,J)=T(L,M,2+K)
4060 J=J+1
4069 NEXT M
4270 I=I+1
4280 IF I>N3 GOTO 4350
4290 IF K=6 GOTO 4320
4300 K=6:J=1
4310 GOTO 3840
4320 L=L+1:K=0
4330 IF L>N GOTO 4350
4340 J=1:GOTO 3840
4350 OPEN3,8,4,"0:WXYZ,SEQ,WRITE"
4360 NO=N3
4370 FOR I=1 TO N3
4380 IF B$(I+1,1)<>"000" GOTO 4400
4390 NO=NO-1
4400 NEXT I
4410 PRINT#3,NO
4420 FOR I=1 TO N3
```

```
4430 IF B$(I+1,1)<>"000" GOTO 4450
4440 GOTO 4510
4450 PRINT#3,D$
4460 BS$(I)=B$(I+1,1)
4470 PRINT#3,BS$(I)
4480 FOR J=1 TO 39
4490 PRINT#3,A(I,J)
4500 NEXT J
4510 NEXT I
4520 CLOSE 3
```

The raw data file (see *Figure 5*) consists of the following.

(i) Date of testing (D$).
(ii) Number of trays (N).
(iii) Blocks of tray results including:
  (a) one line describing tray particulars (T1$);
  (b) eight rows each containing an identifier (A$) and 12 measurements (T(I,X,J)).
(iv) Particulars of strains tested (B$) which include organism code ('org'); strain number (STR$); and results of three tests entered manually (L1,L2,L3).

Each tray contains two sets occupying columns 1–6 and 7–12. Data transformation includes three steps.

(i) Correction for tray/instrument variability by subtracting mean tray values (B1(L)) from each well on that tray.
(ii) Conversion of instrument readings to product units using two different equations (U) for the two types of products present in different positions on the tray (lines 840–1130).
(iii) Subtraction of substrate control values (always occupying the first 6 columns of tray 1) from the remaining sets.

The final part of the program writes the calculated attribute on to a disc file (WXYZ) in a format suitable for the ID program.

10. APPENDIX II

Assignment ID of two OTUs to one of three possible groups on the basis of two continuous and one binary attribute using discriminant function analysis model.

(i) Raw data of two OTUs for three attributes.

OTU	*Attributes*		
	a	*b*	*c*
A	1099	292	0
B	−221	−78	1

(ii) Data after re-coding negative values to 0.1 and taking natural logarithm of values for attributes a and b.

OTU	*Attributes*		
	a	*b*	*c*
A	7.002	5.677	0
B	−2.303	−2.303	1

(iii) Classification function coefficients.

Group	*Attributes*			*Constant*
	a	*b*	*c*	
1	1.894	0.533	0.048	−9.457
2	1.698	0.130	10.194	−9.997
3	0.414	0.390	0.030	−2.259

(iv) Equations for classification scores.

$$CSCR_1 = 1.894*V_a + 0.533*V_b + 0.048*V_c - 9.457$$
$$CSCR_2 = 1.698*V_a + 0.130*V_b + 10.194*V_c - 0.997$$
$$CSCR_3 = 0.414*V_a + 0.390*V_b + 0.030*V_c - 2.259$$

$CSCR_i$ = classification score for group i

V_j = data value for attribute j of OTU_i

(v) Classification scores and choice of group.

OTU	*CSCR for group*		
	1	*2*	*3*
A	<u>6.830</u>	2.616	2.854
B	−14.998	<u>−4.013</u>	−4.081

(vi) Unstandardized canonical discriminant function coefficients for calculating the discriminant score of an OTU on each function.

Discriminant function	*Attributes*			*Constant*
	a	*b*	*c*	
1	0.206	−0.106	3.333	−1.371
2	0.467	0.089	−1.143	−2.815

(vii) Equation for calculation of discriminant scores.

$$DSCR_1 = 0.206*V_a - 0.016*V_b + 3.333*V_c - 1.371$$
$$DSCR_2 = 0.467*V_a + 0.089*V_b - 1.143*V_c - 2.815$$

$DSCR_i$ = discriminant score on function i

(viii) Discriminant scores.

Function	*OTU*	
	A	*B*
1	−0.533	1.733
2	0.961	−5.238

(ix) Group centroids.

Function	*Group* 1	2	3
1	−0.534	1.986	1.482
2	1.053	−0.476	−1.696

(x) Squared distances from centroids of groups on each function.

OTU	*Group* 1	2	3
A	0.008	8.407	7.958
B	44.705	22.749	22.877

(xi) Most likely groups and probabilities of ID.

OTU	*1st group*	*Probabilities* *P(X/G)*	*P(G/X)*	*2nd group*	*Probabilities* *P(X/G)*	*P(G/X)*
A	1	0.996	0.967	3	0.024	0.018
B	2	0.000	0.517	3	0.001	0.483

OTU A is an example of a reliable ID as both P(X/G) and P(G/X) are more than 0.9. OTU B is an example of unacceptable ID, as the P(G/X) suggests that the OTU is only slightly (0.034) nearer to group 2 than to group 3. Moreover, the P(X/G) for either group is very small, suggesting that the OTU belongs to neither.

CHAPTER 5

Computers in teaching microbiology

PETER G.G.MILLER and RON M.ATLAS

1. INTRODUCTION

Computers are influencing education in several ways. In some instances they are used directly in the teaching of microbiology, principally through the use of computer-assisted instruction (CAI, also known as computer-assisted learning or CAL) and laboratory simulation. Such CAI is being used effectively at some institutions to augment traditional teaching methods. Both lecture and practical laboratory coursework can be enhanced by using appropriate CAI packages. However, as with any other teaching aid, critical judgements must be made in the selection of computerized instructional material. In this chapter we will consider some of the practical aspects of using computers to aid in the teaching of microbiology. We will also highlight areas where developments in computer-based educational technology are likely to occur within the next few years. Finally we will address the more general aspects of information technology as they affect students and teachers.

It is important when considering the use of computers in teaching to define clear objectives and the best means of attaining them. To 'computerize' a course simply for the sake of it may be creating needless work. Where existing teaching methods are both effective and efficient there is little point in employing computers except, perhaps, in revision. It is better to identify topics which students find difficult or arcane and which might benefit from the advantages peculiar to computer-based teaching such as animated graphics, the control of pace and direction and so on.

2. IMPLEMENTATION DECISIONS

Questions to be addressed at the start of a particular project must reflect individual circumstances. The impetus for new developments is often provided by junior academic staff. However, if it is likely that their support will be needed at a later date attempts should be made to involve heads of departments, academic and technical staff at an early stage. Those considering investing significant amounts of time in developing software should ensure that this effort will be perceived as a positive factor in their career development. Inevitably many decisions and compromises will need to be made and for large projects a management committee provides a useful forum for debate and the delegation of responsibilities. The latter may include hardware maintenance, software development/acquisition and, where necessary, network management.

2.1 **Budget**

The initial investment will depend on local circumstances and intentions. Note, however, that it has been estimated that a recurrent budget of 30% of the initial purchase cost should be available per annum to support the continued operation of a computer-based teaching project (1). This divides roughly into 10% for a maintenance contract, 10% for purchase of new software, perpherals and consumables and 10% for a 'sink' fund to enable replacement after a nominal 10 year lifespan. Such an ongoing commitment must clearly be borne in mind when the scale of the initial involvement is determined if the usefulness of the project is to be sustained. Where this commitment is not possible consideration should be given to sharing new or existing facilities with other departments. Such arrangements may be beneficial both financially and in encouraging cross-fertilization of academic and educational ideas.

Advice may also be sought from other microbiologists actively involved in this area and organizations such as the Society for General Microbiology (SGM) Computer Users Group and the American Society for Microbiology (ASM) Microcomputer Users Group may be able to help. Attempts have also been made to establish a European Bank of Computer Programs in Biotechnology (2). In the UK the Council for Educational Technology (3 Devonshire St., London W1N 2BA) and the Computers in Teaching Support Service (SWURCC, University of Bath, Claverton Down, Bath, Avon BA2 7AY) may also be useful contacts. Journals such as *Binary, CABIOS* and *Biochemical Education* are sources of specialist ideas and information and journals such as *Computers in Education, CTISS File* and *Academic Computing* deal with wider issues.

The local computer centre should also be contacted at an early stage. It may be willing to defray costs and provide operating staff and invaluable expertise. In any event you should ascertain what level of support they can offer and any favourable terms they might be able to negotiate for hardware, software and maintenance. Equally important they will be able to indicate the feasibility of connecting a system to existing and future campus information services.

2.2 **Microcomputers and operating systems**

Historically much computer-based teaching has used mainframe and minicomputers. The price/performance ratio of microcomputers is so favourable at present, however, that they are likely to dominate this field for the foreseeable future and this chapter deals principally with this type of machine.

It is important to realize that programs designed for use with one type of computer cannot generally be used with another. The widespread use of microcomputers by the commercial sector, however, has led to the establishment of *de facto* standards for the transfer of data and, to some extent, of programs between different machines based on the Intel 80×86/80×88 series of microprocessors and using the MS(Microsoft)-DOS disc operating system. This includes, of course, the IBM PC and PS12 microcomputers and a large number of so-called 'PC-compatibles'. The incorporation of such a standard in more recent machines has involved some compromise in performance and other machines based on the Motorola 680×0 microprocessor series have also become popular. The prime example here is the Apple Macintosh series which employs a user-friendly graphical interface based on Windows/Icons/Menus/Pointers (WIMP) in preference to the command-line driven interface typical of most traditional operating systems such as MS-DOS. The popularity of such interfaces has led to their appearance on MS-DOS machines with packages such as Digital Research's GEM and Microsoft's

MS-Windows together with programming languages such as ACTOR and SMALLTALK. The graphical interface of MS-Windows is also found on the OS/2 operating system for the IBM PS12 series of machines. It seems likely that software in general will increasingly use WIMP interfaces although these can incur appreciable additional overheads in terms of computer memory and development costs. The most significant benefits that derive from environments such as MS-Windows are the ability to execute multiple programs at the same time in separate screen windows and to transfer information between programs.

The continued availability of hardware compatible with existing software is essential in the long term for maximizing the return on software investment. It is most likely to be realized where there is a large user base as with the IBM PC. One consequence of this has been the adaptation of other machines to run some IBM PC software either with software emulators or with the help of additional hardware although levels of compatibility vary, performance may diminish and keyboard differences can also cause problems. The IBM PC, Apple II and Macintosh series, BBC and RML Nimbus microcomputers are other common teaching machines whose manufacturers have attempted to provide some degree of backwards compatibility in new products (the IBM PS12, Apple Macintosh II, IIGS, BBC Master/Archimedes and RML AX/VX series, respectively).

The selection of hardware should be determined by its suitability for the task in hand and the phenomenon of 'gold-plating' (i.e. the purchase of fewer machines with an inappropriately higher performance) should be avoided. From teaching as opposed to research points of view the immediate need is for more software and the performance of machines using true 16-bit microprocessors should be adequate in the medium term. More important is the ability to address substantial amounts of RAM (at least 0.5 Mb) and to have access to large amounts of file store (at least 0.5 Mb).

There may be advantages in purchasing as many items as possible from a single reputable supplier, particularly one with a proven track record in the educational field and the resources and organization to support the end-user. Although this may not involve the least expensive equipment it can encourage discounting and reduce problems which might otherwise arise when items from two different suppliers fail to work together. Where low-cost 'compatibles' are purchased some assurances should be sought on the continued availability of spare parts if these are not directly interchangeable with the authentic machine. In any event it is wise to obtain quotations from multiple sources and to obtain all significant undertakings regarding performance and warranty in writing.

2.2.1 *Local area network versus standalone microcomputers*

In teaching there is often a need to provide the same basic software environment to many machines. The linking of computers through a local area network (LAN), although it adds significantly to costs, can help here and bring about a number of other benefits.

(i) *Sharing of peripherals.* These generally include access to file space on a common disc storage device and use of one or more network printers, access to which is regulated by one machine acting as a printer server. The file server generally controls a hard disc drive which may be most conveniently backed up via a tape streamer. Note that the file server is not available for student use and thus represents an extra expense when compared with standalone machines.

(ii) *Sharing of software*. Files may be readily downloaded to any or all of the machines on the network and the latter may share data or actual programs. Generally users have access to file directories of two types, shared and private, and access to the network is therefore controlled by a user name and associated password. It is important to note that copyright problems may arise in the multiple use of single-user software across a network and where possible network-specific software should be purchased. The acquisition of network-based software may bring appreciable discounts although some suppliers remain dubious about networks.

(iii) *Messaging*. Some systems support the transfer of files from the file space of one user to that of another, others will actually cause text to be displayed on another machine or can permit the monitoring of activity on another computer as on the Acorn ECONET system.

Other benefits should include automatic time and date stamping of files. LANs also permit the operation of discless stations which depend on the file server for their file store. Such machines are obviously cheaper and eliminate problems of illegal copying of software but they do put extra strain on the file server and local discs are probably worth maintaining as they enable off-line use, provide potentially secure file storage and place at least some of the onus on the user to manage his own files. The LAN should also incorporate software support for a network manager to assist in mounting software, maintaining bulletin boards, assigning passwords and privileges, controlling the proliferation of files and diagnosing and correcting faults. Problems, many trivial and some imaginary, will inevitably arise and some formal arrangement must exist for their solution. The network manager should act as a point of reference for such problems.

One alternative to a LAN is a multi-user system in which terminals are serviced by one CPU. Predictably this results in performance that is very much dependent on usage and some types of common software may not run in this environment.

It is advisable, though not always possible, to test a chosen LAN or multi-user system of the same size and operating under the conditions you anticipate before making any commitment to purchase the system.

2.2.2 *Peripherals*

A range of additional items may also be needed. Some, such as monitors, are essential. A question to be decided is whether the additional versatility of colour monitors justifies the extra expense. This will depend upon what use, if any, the software makes of colour. Used properly colour can do much to enhance teaching software although it is important to check that the monitor supports legible character fonts as well. Some monitors have anti-glare coatings and tilt/swivel mechanisms although siting should also be a factor in solving ergonomic problems.

Another important purchase is a printer. Various types are available and important factors, particularly where network use is anticipated, include robustness, economy (both in purchase and running costs), ease of operation and speed. A dot matrix printer with draft, near letter quality (NLQ) and graphics modes is generally a safe choice although such printers tend to be noisy. It should obviously support the intended software although the use of resident 'screen grabbing' software, as supplied, for example, in

MS-Windows, can be useful for capturing screen output from some non-supported software and sending it to the printer.

Although the quality of graphics printing has improved with the advent of 24-pin dot matrix printers the jaggedness typically seen in diagonal lines is best avoided by the use of a plotter or laser printer. Low-cost plotters typically require considerable attention during operation and cannot cope with word processing requirements. On the other hand laser printers, with appropriate software, cope admirably with both tasks although they can be expensive to run. The more expensive laser printers generally support a page description language such as PostScript and contain their own memory and microprocessor to interpret the print file commands. This arrangement lessens the burden on the host microcomputer and generally affords a superior result.

2.2.3 *How many machines are needed?*

The ratio of machines to students and the number of courses involving computers will obviously influence the pattern of use. There is frequently some merit in pairs of students using each computer as they will generally learn from one another as well as from the software although there is a risk that one keen student may effectively exclude a less adept partner. It is not realistic to develop computer-intensive courseware for a large class when only a small number of machines are available. On the other hand where use is likely to be low, intermittent or designed to concentrate on tutor-led demonstrations, then a smaller number of machines will be adequate. It should be noted, however, that if computer use is intended to fall within the framework of an existing timetable then competition between courses for access to machines may not be distributed evenly across the working week. If students are given free access and expected to complete assignments on the computers in their own time then their needs must also be taken into account in timetabling. Whatever the local arrangement it is essential to provide time in the curriculum for the use of computers and this may entail rescheduling other activities.

Some US universities now require that freshmen students purchase their own computers on arrival and negotiate substantial discounts with suppliers on this basis. In the UK the Nelson report envisaged a less ambitious ratio of one workstation per five students by the year 2000 and the UGC has granted £11 million to establish more than 100 pilot schemes in departments throughout the UK (3). The onus will then fall on the universities to continue the work after the grant period expires.

2.2.4 *Where should teaching computers be located?*

The temptation is to place all the machines in a single room. This makes supervision easier and, where networking is planned, cables can be surface-mounted and costs reduced considerably compared with the cost of installing cables over any appreciable distance through partitions or between floors. Fault-finding may also be simpler. There is, however, much to be said for putting some machines into laboratories and offices which at present lack them. If they are supplied with word processing software, for example, this can have the effect of generating a cadre of interested academic users who are competent at operating the machines and more sympathetic to their application in teaching. Such machines are also more accessible to students undertaking laboratory-based project work. In general there is no greater disincentive to using a computer than having to walk an appreciable distance to use it.

There is also a case for having a cluster of machines for small group teaching. Specialized facilities such as large monitors, printers, plotters and so on can then be provided for such groups. A recent innovation useful in tutorials is a liquid crystal display that can be mounted on an overhead projector, thus permitting the computer output to be projected onto a screen at a low cost relative to dedicated projectors. In general the arrangement of machines within the room should be carefully considered so as to ensure good glare-free lighting and to enable all the students to see the tutor at the same time. Space should also be provided between machines for students to write or use texts. This space is particularly important where mouse-driven software is being used and in such cases it may be advisable to leave space to the left as well as the right of the machine and to ensure that the surface of the desk or bench is suitable for mouse traction. It should be possible for the tutor to circulate freely amongst the machines, to observe the students, comment on their progress and answer their questions. It is a common observation that students interact more readily with their tutor during such sessions although this may be more an indictment of standard tutorial formats than a specific virtue of using computers. It also follows from this observation that such tutorials do not constitute a rest period for academic staff!

2.2.5 *Physical security*

Microcomputers are valuable items and as such are increasingly prone to theft. It is wise to take this into account at an early stage, for example by consulting with local security specialists if possible. A number of devices can be purchased or fabricated which will render machines immovable or unusable or sound an alarm if the machine is moved. The choice will depend on make of machine, cost and factors such as anticipated frequency of relocation. It is essential that such devices should not impair the normal use of the machine, including the use of peripherals. In any case it is a sensible precaution to indelibly mark all the machines and their associated peripherals. Whether the agglomeration of machines into a single room makes a bigger target or renders them easier to protect is debatable.

2.2.6 *File security*

Risks associated with file security operate at two levels, firstly the inadvertent loss of a file and secondly unauthorized access to other users' files. In the first instance one must stress the need for users to maintain their own back-up copies of important personal files on local floppy discs. A convenient source of such discs will need to be arranged if none exists. As a second line of defence network managers should ideally make copies of the file server winchester disc onto some kind of removable medium, typically magnetic tape. This requires that some routine be established for backing-up, particularly in cases where the network cannot operate during this process. Ideally multiple copies should be maintained in distinct geographical locations.

Where a network is used routinely by students it is wise to discourage staff from using the network file server for storing confidential records and in general staff should be made aware of the legal position concerning such records. Most commercial networks are accessed via user names with associated passwords. This provides a modicum of security though users are notoriously unimaginative in their choice of passwords. In some systems the file server is a weak link in security and it is often a good idea to keep it locked in a separate room. Many staff are not aware that under some operating systems, MS-DOS for example, files are not finally erased until they are overwritten and until this occurs may be 'un-erased' by suitable software. There may also be dangers in programs that create intermediate working files or automatic back-up copies of earlier versions of files.

Additional data privacy may be conferred by encrypting files, hiding filenames so that they do not appear in standard directory listings and, on networks, using low-level calls to the network operating system to save data to a directory or user name that students cannot normally access. It may also be possible to log the time of entry to and exit from programs in order to guard against multiple attempts at tests, for example, and in general it is a good idea to prevent students breaking out of programs and into the operating system. If the latter course is followed, however, special provision must be made to enable students to manage their files.

Problems may also result from students deleting files wanted by others to make space for their own. Most networks will permit the protection of such shared files by assigning a certain level of privilege, variants of read-only/write-only/create, to a user, thus limiting the scope for saving files. It may be possible to restrict access to the file locking utility (ATTRIB.EXE under MS-DOS) in order to prevent the unauthorized unlocking and deletion of files.

3. ASPECTS OF COMPUTER USE IN TEACHING MICROBIOLOGY

3.1 The computer as a tool

One critic of computer use in education has argued that the advent of the word processor, spreadsheet and personal database constitute such a significant advance for any one invention that to expect additional benefits is unreasonable and further claims should therefore be moderated! (4). The central role that computers are playing in many aspects of administration has led to the belief that students should become conversant with 'information technology'. This stresses the function of the computer as a general-purpose tool, a function quite distinct from its use in learning. Features that distinguish a good tool are its ease of use, its efficiency and its flexibility. Ease of use is engendered if a program has a consistent, intuitive and forgiving user interface and it is this feature which has highlighted WIMP software on machines such as the Apple Macintosh. Efficiency is expressed as economy of effort in achieving the desired goal, thus increasing speed and productivity. Note that selection of a task by clicking on an icon in a WIMP environment may be efficient but typing out a word may have educational merit in terms of reinforcement. Finally a good tool is essentially content-free: it can be used for a range of functions, either well or badly. The user will wish to apply his tools with the least amount of interference due to actual context—hence the popularity of multi-tasking environments which allow multiple programs on screen at the same time and permit data transfer between them.

3.1.1 *General tools*

Most of the market leaders for various general applications are well known and have been extensively reviewed in the popular computing press. In many cases, however, much less expensive 'clones' are also available on the market which provide much the same degree of function. In any case it is necessary to be sure that the level of sophistication required to run a business is equally appropriate in teaching. Sophisticated applications require more training although, of course, this training may have a longer term benefit for the student. In general WIMP-based systems provide a more consistent and thus more readily assimilated software environment. Such consistency is also generally afforded by 'integrated' software suites covering a range of applications although the individual elements may be less powerful than their counterparts marketed as separate products.

3.1.2 *Word processing*

A wide range of programs are available for this purpose. Students, of course, are required to present a large amount of written work and, like their mentors, appreciate the flexibility of the medium and the quality of the output. There is a tendency to see word processing as an abuse of computers earmarked for teaching. Quite the opposite is true: it is a clear benefit of the technology and provides an excellent opportunity to practise keyboard skills. A balance needs to be struck, however, as it is a computer-intensive activity that can become a drain on resources especially if undertaken by two-finger typists composing their text on-line. Some word processors also provide thesaurus and spelling-checker tools which can be used by the less gifted writer to improve the quality of his work. It is important to remember that scientific word processing has its own peculiar requirements such as Greek and symbol fonts and that the average word processing package and printer may provide minimal support for such functions.

3.1.3 *Spreadsheets*

Microbiology students do not in general have extensive repetitive calculations to carry out or 'what-if' projections to perform and the majority will already possess a sophisticated pocket calculator adequate for their computational needs. Spreadsheets have been used for a variety of simulations but perhaps they are best employed in their most obvious role, to cost some enterprise, perhaps a biotechnological process. Clearly it is facile for students to use a prepared spreadsheet but it is more instructive if they build their own. One problem with such spreadsheet modelling, however, is that the onus is placed on the student to flag an absurd answer. It can be useful if a spreadsheet also includes graph plotting capabilities and some offer additional functions such as a macro language, project management aids and elementary database functions.

3.1.4 *Databases*

The primary areas in which databases are encountered by students are in bacterial identification and in the maintenance of bibliographies. The results from test kits such as the API 2OE system may be most readily analysed by microcomputer. On-line bibliographic databases are also available but are generally too expensive to allow students to access except under the supervision of trained library staff. Libraries are themselves becoming more highly automated and students should be made aware of any in-house computer facilities for searching the catalogue, determining availability and making reservations. Students may also construct a database of their own references although it is doubtful whether this ought to replace the more traditional methods at undergraduate level. It is important to stress the appropriate and inappropriate use of computers.

Where networking is available, controlled access to records in the same database may be particularly useful and is a feature of some spreadsheet and database packages. Multi-user programs are particularly useful for data collation in practical classes and for running student questionnaires or tests. Generally, however, software intended for single machines will enable only one user at a time to edit a file across a network and some software may not run at all on a network, particularly if it employs certain types of copy protection.

Videotext-like databases can be mounted on standalone or networked systems as public information systems. However, the anticipated benefits should be compared with those accruing form a well-organized noticeboard before such a system is installed.

3.1.5 *Other applications*

Certain departments may have a need for statistical functions over and above those offered by spreadsheets. In addition to a wide range of commercial and public domain packages a book has recently appeared with listings in Pascal (5). Graph plotting and curve-fitting programs may also be purchased commercially but will in general be hardware-dependent.

3.2 **Languages**

Whether the ability to program in a high level language constitutes a part of computer literacy is a moot point. Such a skill could obviously be acquired in a service course taught by specialists. What does seem clear is that there is little point in gaining proficiency in a language without providing the opportunity to practise it and that this will be a function of local attitude and resources. Traditionally the language of mainframe scientific computing has been FORTRAN and this is particularly true where access to libraries of numerical methods is required. The transition to microcomputing, however, has led to the widespread use of other languages, generally simpler than FORTRAN, such as BASIC, Pascal and LOGO. Each language has its proponents and there appears to be a general convergence on the standards of program structure established by Pascal.

The most significant language developed specifically for education in recent years is LOGO. This was originally intended as a language that was rewarding visually and that encouraged structured programming and, as a consequence, analytical thought (6). It is commonly used in schools but is rarely encountered in tertiary education. Its drawbacks include speed of execution and lack of standardization. However, where only a limited amount of programming is anticipated its simplicity may have advantages, especially where students have encountered the language previously.

3.3 Scientific tools

The first microcomputer in a department is often purchased for a scientific purpose like data-handling or nucleic acid sequence analysis. Much software is available from commercial sources and in the public domain for such needs and some is described in other chapters of this book and in other volumes of this series. A listing of biochemical software is provided by (7). The interplay between teaching and research that distinguishes tertiary education suggests that such software should be amenable to teaching use. It is, however, important to be aware of the complexity of some of these programs and their idiosyncratic user interfaces and documentation. Clear guides to such programs would be useful companions for students. Use of such tools should not obscure the fact that the problem to which they are being applied should have educational validity. Such programs do not normally go out of their way to be visually stimulating and their use on a boring problem soon leads to disaffection amongst students.

3.4 Computerized test banks

Another indirect effect of computers on teaching is the development and use of computerized student evaluation systems. Some text books now come with computerized test banks that are intended for use by course instructors. Test questions are generally of the multiple choice type. Tests can be prepared very rapidly, allowing instructors to concentrate on teaching rather than on examination preparation. When used in conjunction with computerized grading systems, tests can be graded, recorded, analysed and returned to students within minutes rather than days. Such computerized test systems are particularly useful when teaching large classes and can provide the instructor with rapid feedback instead of or as well as formal assessment procedures. Concern, however, must be raised that the exclusive use of such tests cannot help to develop the analytical and synthetic skills needed for success in essay-based examinations.

Besides test banks intended for instructor use, there are others for student self-evaluation. These are often produced using an authoring software applications package which requires only the entry of the question and answer; the authoring package provides the program needed to generate the exam and the corresponding key. In generating such test banks, record keeping decisions must be made; should the instructor have access to the student's test performance or should the student's privacy be maintained? In some cases instructors prepare tests to match specific course content whilst others may be of a more general nature. An example of an effective test system for student use has been produced by the Uniform Services University. It is intended to help prepare students for medical board examinations. This test bank covers many areas. The system includes a manager that permits editing or the addition and deletion of questions so that they may be updated as needed.

3.5 Laboratory teaching

In an experimental discipline such as microbiology practical classes have a major part to play. Ideally they should reinforce lecture material, teach manipulative skills, allow the practice of experimental design and the analysis of data. In terms of labour and materials they are significantly more costly than lectures and it is therefore pertinent to ask whether computers can be used to make practical sessions more effective. Possible functions include the following.

(i) Database and news system to provide students with background information and updates to schedules.
(ii) Data collation for the comparison of results obtained by different students, perhaps using the same test with different species.
(iii) Data analysis, including elementary statistics and curve-fitting.
(iv) Simulations of experiments that are too costly, dangerous or time-consuming to mount. Such programs may involve the variation of parameters with a view to optimizing a particular process as with Bungay's FERMT penicillin fermentation (8). Where expertise exists students could be encouraged to develop and explore their own models. The availability of modelling environments such as MicroModeller and Academic Stella may encourage this type of activity in departments unwilling or unable to devote time to programming. The results are inevitably somewhat divorced from the 'real' world although some packages, such as the chemostat model CCSP, introduce random 'noise' to relect more closely data that might be generated in the laboratory (9).
(v) Data capture using analogue or digital interfaces to equipment such as spectrophotometers or fermenters (10).

3.6 **Computer-assisted instruction**

A number of CAI packages have been developed to help in the teaching of microbiology. In some cases the programs are simply electronic study guides, that is, they use a computer to present information on a video screen. Such 'page-turning' texts are easy to write especially with authoring systems. Although they undoubtedly have their place such packages in general do not use the computer to maximum effect and may not be especially cost-effective. Indeed they often compare unfavourably with books which naturally afford features like random access and annotation. Attempts to improve the conventional word processor have resulted in the concept of 'hypertext' which goes beyond the normal serial text format by allowing words on screen to be used as keys, for example, to a definition or explanatory graphic or to act as an entry point to a deeper level of explanation.

The better CAI programs take advantage of the power of the computer and can interact with the user. They can include animated graphics and interactive logic branching that cannot be readily achieved with other media. These programs are best suited to problem solving, particularly when new versions of the problem can be randomly presented to the user and when the user can follow various paths, sometimes with the aid of computer-prompted suggestions, to find the solution to the problem. An important component of such systems is the ability to analyse student responses and to act accordingly, perhaps by congratulating the student or by giving him a hint before repeating the question.

3.6.1 *Writing software programs*

The writing of software programs for CAI can be achieved in several different ways. Some microbiologists are competent computer programmers and can combine their knowledge of microbiology with their abilities to write programs. The relative abundance of programs dealing with aspects of growth kinetics may reflect the programming capabilities of academics in this field as much as the amenability of the subject to CAI. More often, however, microbiologists are not expert computer programmers and, in these cases, it is best to rely upon a team that includes a microbiologist and a computer programmer in order to develop appropriate software. Within a writing team an understanding must be developed as to the limitations of computers and the difficulties in achieving certain types of operations. Microbiologists must work within the limitations of the screen size of the particular computer system that is chosen as a minimal hardware requirement. Often there must be extensive editing to maximize the information presented on a particular screen. Microbiologists must also become aware that small changes can require great efforts on the part of the computer programmer. This is particularly true for the presentation of graphics and for changes in the logic sequence paths that may be followed through the program. It is clearly advantageous to define the function and appearance of the product before coding actually begins.

There are computer authoring packages available that can assist in the development of microbiological computer software of the tutorial genre (11). Such packages can greatly speed the development of CAI software but they tend to be expensive and their capabilities inevitably represent only a subset of what is possible with a conventional programming language, particularly with regard to graphics. Their principal advantages lie in the provision of routines to analyse student response and to control branching, thereby allowing the academic to focus on the educational objectives at an early stage. It should be noted, however, that writing sophisticated teaching applications with such systems is likely to come easier to persons with some experience of programming: in the final analysis they themselves represent a simplified, high-level language.

Software written for use with one particular type of computer can, of course, be translated for use with another type of computer. However, this is often much more difficult than would at first appear. In cases where the initial program is written in BASIC or a low level programming language and/or where extensive graphics are employed, the translation process can be particularly difficult. Often it is best simply to determine what the program is aimed at achieving and to initiate the writing of software for another computer system as if it was an entirely new programming effort and this time incorporating programming practices likely to enhance portability.

Programs that use a minimum of read or write to disc operations are best suited for teaching purposes because of the relative slowness of carrying out such operations. It is usually best to load into memory the entire program so that no disc operations are needed. Judicious editing of a program may be needed with some machines to permit limiting its size to that which can be stored in the computer memory.

Ideally the program should be 'idiot proof', that is the user should simply insert the program diskette and follow the instructions printed on the screen. The instructions should be simple commands, such as 'Press the return key' or 'Enter a letter corresponding to your choice', or 'Menu option number'. No matter what the user does the program should have an appropriate response. If the user presses an incorrect key, a prompt message should appear that again specifies the limitations of the appropriate response. The CAI program must not itself add a level of confusion to the material being presented.

A good program should be adaptable to its user's preferences. A program written for a colour monitor should also be operable on a monochrome screen and ideally the selection of colours used should be capable of modification by the student, perhaps via an optional configuration file. Colour should be used with discretion and preferably to impart extra information rather than solely for effect. Thus help screens or keywords could be highlighted or shown in a different colour or font from standard text to attract attention and identify their purpose. The use of such colour coding should, of course, be consistent. Be wary of excessive use of sound to flag error—this can be annoying to others and embarrassing to the individual involved (although it does have the advantage of identifying him readily). Group common actions together in menus but try to avoid multilevel menus in which it is easy to get lost. If these are necessary, provide a printed route map to help novices find their way around. Remember that interest will decline if execution is unreasonably delayed—where delays of longer than 5 sec are inevitable always indicate that something is happening or the student will probably assume the program has crashed. Observe students using your software—most students are remarkably uninhibited about criticizing deficiencies. Additional guidelines for the design of user interfaces can be found in (12,13).

3.6.2 *Proofing and de-bugging programs*

It is critical that CAI programs be carefully proofread and that all possible paths through the program be examined. The utility of CAI packages is badly compromised when typographical or programming errors occur. Unlike the proofreading of hard-copy material, the proofreading and de-bugging of software requires that the screens be viewed and the numerous branches of the program be tried whenever any changes are made. It is insufficient just to examine the section of a program where a change is made since changes in one section can have knock-on effects, altering other parts of the program. Thus proofreading and de-bugging programs to ensure accuracy must be a major part of the development process and adequate time must be devoted to it.

3.6.3 *Software logic*

Various logical sequences can be employed as a framework for CAI packages. In its simplest form the program may be unidirectional with no ability to branch. A slight modification of this logic is to permit bidirectional movement through the program, that is the user may proceed in the forward or the backward (review) mode. Beyond this the program may become more complicated with branches at various points so that the user may select the direction in which to proceed. In some programs branches at one point eventually lead back to that same point, that is subroutines from a particular point in the main program lead back to the same point in that program. Even more complex interactive programs permit a wide variety of branch selections.

Besides the logical flow pathway of the program, it is necessary to make various other logic operator decisions. For example, when asking a question of the users, should the users have the option of skipping around that question and seeing the answer or should they be permitted to proceed with the program if they do not know the answer? Some programs allow students to skip sections once they have demonstrated a mastery of the topic. Should there be portions of the program that are considered more advanced and hidden from the user until they have completed particular sections of the program?

If the user has difficulty with a particular concept being presented, should they be allowed to branch off into additional remedial tutorials before continuing with the material in question? These are the sorts of logic decisions that must be thoughtfully made when developing software for computer-assisted microbiological instructors.

3.7 Use of CAI packages

There are various ways in which CAI packages may be employed. Thus they can be used to present information in a way that makes it easier to remember. Active graphics and sound effects can impart information to the user in a way that is easy to visualize and recall at a later time. The presentation of information in a 'games' format can make learning so enjoyable that the user rapidly masters knowledge of the subject. Another excellent way of teaching material is to present problems and to allow for interactive responses; if the user branches in an inappropriate direction the computer can offer prompts that will place him back on a correct path to solving the problem. The design of the program must anticipate virtually all potential responses and provide the necessary prompts that will allow the user to reach an appropriate point in the program.

3.7.1 *Selection of software*

A common question that people will ask about a new system is 'what will it do?' and from a strategic point of view it is a good idea to ensure that software is available that they will feel motivated to use from a very early date. Word processing software and, to a lesser extent, games are useful acquisitions in this category.

Many programs are 'homegrown', that is, they are developed by individuals for their own use at their own institutions. Such software is often free but its quality variable. Often it is intended to meet a specific need and is not developed to the point that it can be used effectively by others. Finding such software is often fortuitous although the SGM Computer Users Group maintains a catalogue which includes a number of programs which may be useful in teaching and a general review of US teaching software sources has appeared recently (14). Demonstrations of software at meetings of the SGM Computer Users Group and the ASM Microcomputer Users Group are additional important routes through which information about such programs is disseminated.

Some software for microbiology is commercially produced and more widely marketed. Such software is often quite expensive and its use may require site licensing. Usually, but not always, commercially produced software is extensively developed and its general applicability tested before it is marketed. There has, however, been a lack of software reviews of such products. Thus, users are often purchasing programs based on marketing literature rather than critically reviewed evaluation. Fortunately some commercial producers of software provide disabled or demonstration programs or allow return of material should it not meet the buyer's expectations. Greater use of critical reviews, such as those that appear in the journal *Binary*, is essential for guiding microbiology instructors to appropriate CAI packages.

3.7.2 *Examples of computer-assisted instructional packages*

A list of some of the software available for use in teaching microbiology is given in *Table 1*. A small number of examples will be discussed in somewhat greater detail to

Table 1. Software for teaching microbiology.

Topic	*Program*	*Category (machine)*	*Source (author)*
Authoring systems	Course-builder	U(M)	TeleRobotics
	Microtext	U(I,B,N)	Transdata
	TopClass	U(I,N)	Format PC
Hypertext	Guide	U(I,N,M)	OWL
	HyperCard	U(M)	Apple dealers
Menu generator	MenuGen	U(I)	Microsoft
Evaluation	CourseView	U(I)	CourseView
Medical microbiology	Micro-Micro	CS(A)	Macmillan (Snyder, Weinberg, Atlas)
	Mical	CS(I)	Upjohn (Levinson, Gardner)
Taxonomy and identification	Bacterial identification	SG(I)	IRL Press (Bryant)
	NUMTAX (numerical taxonomy)	U(A)	SGM (Colby)
Enumeration	Plate count practice	S(B)	Ref. 15 (Stewart)
	Most Probable Number	S(B)	Ref. 16 (Stewart)
Growth kinetics	PLASPLOT S(B)		Ref. 17 (Cooper)
	OXYDEN (oxygen transfer/ uptake)	S(B)	Ref. 18 (Schaffer)
	BATCHSIM	S(B)	SGM (Prosser)
	Chemostat model	S(A)	SGM (Colby)
	FERMEN, FERMT	S(B)	Ref. 8, SGM (Bungay)
	CCSP (Chemostat model)	S(BASIC, FORTRAN)	SGM (Bazin, Saunders, Prosser)
Microbial genetics	GENESIM	S(IBA)	BIOSOFT (Wood)
Modelling	MicroModeller	U(I,B)	IRL (Salmon)
	Stella	U(M)	Logotron

Only programs mentioned in the text or implemented on commonly used teaching machines are listed. Key: Category: C, CAI; S, simulation; G, game; U, utility. Machine: I, IBM PC or compatible; B, BBC microcomputer; A, Apple II; M, Apple Macintosh; N, RML Nimbus.

Sources: CourseView Ltd, Northgate House, St Mary's Place, Newcastle-upon-Tyne NE1 7PN, UK
BIOSOFT, PO Box 98, Cambridge CB2 1LB, UK/PO Box 580, Miltown, NJ 08850, USA
Format PC Ltd, Goods Wharf, Goods Road, Belper, Derbyshire DE5 1UU, UK
IRL Press Ltd, PO Box 1, Eynsham, Oxford OX8 1JJ, UK/IRL Press Inc., PO Box Q, McLean, VA 22101-0850, USA
Logotron Ltd, Dale's Brewery, Gwydir St., Cambridge CB1 2LJ, UK
Macmillan Publishing Company, 866 Third Avenue, New York, NY 10022, USA
Microsoft Technology Ltd, The Old Powerhouse, Kew Gardens Station, Kew Surrey TW9 3PS, UK
OWL International Inc., 14218 Northeast 21st St, Bellevue, WA 98007, USA
OWL Ltd, Rosebank House, 144 Broughton Road, Edinburgh EU7 4LE, UK.
SGM: contact the author or the Convenor of the SGM Computer Users Group, c/o Harvest House, 62 London Road, Reading RG1 5AS, UK, for details
TeleRobotics International Inc., 8410 Oak Ridge Highway, Knoxville, TN 37931, USA
Transdata Ltd, 11 South St., Havant, Hants PO9 1BU, UK.

outline the ways in which CAI packages can aid in the teaching of microbiology.

(i) *Micro-Micro*. Micro-Micro is a package of 42 lessons emphasizing medical microbiology and is available for the use with the Apple II and IBM PC microcomputers. They were developed for use with undergraduate classes. The lessons are of various types and intended for different uses.

A number of programs present tutorial information, for example the tests used in diagnosis of urinary tract infections. This is followed by a series of multiple choice questions. These programs contain limited graphics presentations and are intended for use by students with specialized interests. Specifically they are aimed at students seeking information and knowledge of a particular area that is not necessarily within the mainstream of general course material. They permit the instructor to refer the student to the computer laboratory rather than taking a great deal of their time to meet the needs of specific students. These programs are intended for one time use by a particular student.

There are several general programs in Micro-Micro that make more effective use of the power of the computer. For example, there is one lesson dealing with enumeration methods that presents graphics of serial dilution and plate count techniques and requires the student to complete a particular level of task before proceeding to a higher level. This lesson is particularly well suited for use with a computer because the inclusion of a random number generator within the program means that an infinite number of different problems can be set. The program contains the correct algorithm for determining the answer to a particular problem and the computer checks the student's response against the computer-generated answer. In fact CAI programs are most at home when dealing with mathematical problems of this type.

In other programs dealing with growth and death of microorganisms the student is provided with an option for graphing the results specified by particular parameters or for using the computer as a mathematical calculator in order to determine the answer to a particular problem. In some of these programs appropriate prompts are used when a student gives an incorrect answer, such as, 'That number is too high', or, 'too low'. In this way the student is directed toward the appropriate answer.

In Micro-Micro, questions to which the student has given an incorrect answer are stored and at the end of the program the user is allowed to review those questions they missed on the first attempt. They are also given a score so that they can judge their own performance but no record is maintained of this score. This is intended to lower the anxiety inherent in taking a computerized test.

In a controlled experiment designed to determine the effectiveness of the enumeration program, one group of students was allowed to use the CAI package whilst another was not. Performance by students using the CAI package was 30% higher than in the control group. Thus, at least in this case, the use of CAI was demonstrated to be highly effective in improving student performance.

(ii) *Mical*. Mical is a series of programs intended for use with an IBM PC at medical schools to help future physicians in the development of skills needed to diagnose diseases. At present there are a limited number of case histories but this number should grow in the near future. The programs are produced by Upjohn Company and distributed gratis to selected medical schools.

Mical contains highly interactive programs that allow the user to select the various types of information that go into diagnosing disease conditions. The user selects the order of information that goes into the diagnosis and can obtain information about the previous history of the individual and the physical signs of the disease condition. Additionally, the user can request laboratory tests. Thus, the user is able to obtain the necessary information for making a proper diagnosis and selecting the appropriate treatment as a physician would do but without the need for dealing with actual patients. The program stores the information that has already been gained and allows the user to review this at any time. Users are guided through the programs when necessary by use of a help function with appropriate prompt messages. The only limitation of this program is that each case has a single diagnosis and hence each case can be examined effectively only once. With a large enough library of cases, this problem would be minimal.

(iii) *Growth*. This tutorial-style program deals with bacterial growth kinetics and was written using the authoring language STAF2. It is notable for having been extensively evaluated, the general conclusions being that computer-taught students performed as well as and in some respects better than those taught in a conventional fashion in a subsequent assessment (19). The program text makes a considerable effort to mimic the teaching style of its author and the teaching method was generally well-received by the students with the reservation that an associated printed handout was also regarded as important (20). This and other programs for the teaching of growth kinetics were reviewed by Wimpenny (21).

(iv) *MicroModeller*. This commercial program can be used for the modelling of growth kinetics although it could equally be used in other similar contexts and comes with a number of examples. It allows the user to enter a formula for his model, applies standard numerical alogrithms to derive a solution and then plots a graph of the function.

(v) *Bacterial identification*. This commercial program is both a simulation and an example of the games genre of teaching program. The student competes against the computer to identify an unknown bacterial species by means of a panel of standard tests. The program allows the student to consult an 'expert' (the computer in another guise) to see if his reasoning is good: he is only allowed to attempt a positive identification once he has significantly narrowed the range of possible choices. In the IBM PC version the data matrices can be edited by the instructor according to local needs.

4. PORTABILITY

4.1 **Software portability**

Program development is a labour-intensive and hence costly exercise and it is therefore wise to try to anticipate future developments beyond the current generation of hardware. It is therefore a good idea to choose a language that is portable and to impose the self-discipline of writing programs that access the hardware (screen, disc drives, keyboard) via calls to intermediate device drivers, perhaps maintained in a separate library. Under these circumstances only the drivers need to be rewritten when the code is ported to a new machine. Such calls are inherent in writing for environments such as MS-Windows although this is not a trivial task. Thought must also be given to program documentation so that it remains possible to decipher source code at a later date. To this end languages such as Pascal that support meaningful identifiers and are to some extent self-documenting have much to recommend them.

Another problem with programs aimed even at only one computer type is that there may be various models and clones of that computer. Instructional packages intended for public or commercial use on the Apple II or BBC systems should be able to run on all versions of that system, including variants of the operating system or BASIC. Programs for the IBM PC must not only run on the various versions of that computer produced by IBM but also on the numerous 'clones' produced by other companies. To permit this it is necessary to compile the programs with special attention given to graphics and sound effects. The necessity for extensive professional computer programmer time in the writing of these programs adds significantly to the cost of such programs.

4.2 **Academic portability**

Software authors should attempt to anticipate the likelihood that the usage of their program will change with time. Wherever possible tables of data should be loaded from disc files rather than being embedded in the source code so that they can be readily modified without having to alter and recompile the program. This also aids in the implementation of teaching software in other institutions that may follow a different academic regimen.

5. USER EDUCATION

In many institutions total familiarity with computers cannot be assumed in either students or staff and efforts must be made to educate them in the use of local facilities. They will need to know how to switch on, log-on (for a network), use the keyboard, discs, disc drive and printer before they can tackle a major software package. The keyboard can pose problems, particularly where use of the computers is only occasional, and it is worthwhile writing or acquiring a keyboard tutor program that can be used for revision. This should demonstrate the use of the RETURN and shift keys, functions accessed via other key combinations, cursor keys, delete and destructive backspace keys. The format for the entry of numbers should be described and the distinction between the number 0 and the letter O should be stressed.

On a network it is worth considering how students will locate and execute programs, that is the nature of the 'front-end'. Files on networks, and hard discs in general, tend to be organized into sub-directories with common themes according to a hierarchical or 'tree' structure. Although utilities are available which clarify such structuring they can be avoided by employing a menu-driven turnkey system in which the student selects the software required from a menu and is returned to the same menu when the program ends. Programs are available that go some way towards automating such menu construction, for example MenuGen.

Ideally the form of a program should so clearly mirror its function that its use is transparent to even the novice user. Inevitably, however, large software applications such as word processors, spreadsheets and databases will require additional support. This may include courses, on-line tutorials or help screens or simply the provision of a manual or other documentation. In any case the process requires some thought. It may be necessary to arrange some form of library system for the loan of manuals, particularly on large networks where the users may outnumber the (expensive) manuals.

6. EVALUATION

There has been relatively little systematic assessment of the use of computers in teaching microbiology although the investment required to initiate computer-based teaching has highlighted the role of formal evaluation (22). In truth evaluation is not to be taken lightly and many academics do not have the time to accept yet another burden. The process of pre- and post-testing control and test groups can, of course, be applied to any comparison of teaching methods although it rarely is in tertiary education. Perhaps the ideal situation will arise when formative evaluation is built into the development of commercial teaching packages. CourseView, a database program specifically intended for the evaluation of courses, has recently appeared and a more general critique of the current generation of CAI packages has been made by Self (23).

7. FUTURE DEVELOPMENTS

The immediate need is for more software of a higher standard and addressing the real problems facing microbiology teachers. The software must be error-free, attractive, portable and affordable in bulk. It is possible that developments in authoring systems may make use of increasingly powerful hardware to overcome some of their present limitations and permit the rapid production of a new generation of software. Research in this area is also addressing the learning process itself so that future so-called 'intelligent' CAI programs will be able to build a model of the student and tailor his learning process according to his needs (24).

Hardware developments will continue and techniques such as natural language interfaces and even speech input will become practicable. Microcomputers will take their place as controllers of access to vast databases on CD-ROM or interactive video. The question of the content, purpose and producer of such systems remains to be determined, let alone their market.

8. ADMINISTRATIVE USES

The use of computers to generate the textbooks used in teaching microbiology has improved the accuracy and timeliness of materials presented to students. Word processing programs simplify the revision of manuscripts, allowing authors to respond effectively to reviews and new information. Additionally, the development of desktop publishing software packages enables instructors to prepare their own material to a high standard at low cost and with relative ease. Thus, instructors can prepare lecture note supplements and practical laboratory exercises containing both text and illustrations that meet their specific course needs. The use of programs such as MS-Windows coupled with various desktop publishing and word processing programs and scanners, results in the production of professional appearing products. The result is the improved communication of information.

The use of electronic media for transmission of the manuscripts to typesetters also affects teaching by reducing the time needed to produce the final book. Authors have greater control over the manuscripts and can correct and update information almost until the work appears. Electronic transmission of manuscripts also reduces the inevitable typing errors and simplifies the job of proofreading. As a result, publishers are now able to produce revisions of texts within a 2–3 year period rather than the typical 4–5 year period. Consequently, instructors must revise their lecture notes more frequently in order to have them correspond to the new texts. Students benefit because they see the latest information, which is especially critical in microbiology with its rapid developments of new information in numerous diverse fields.

Spreadsheets and databases may also have a significant role to play in departmental administration, for example compilation of examination results, maintenance of accounts and inventories. It is important to be sure, however, that the effort required to implement such systems is justified in terms of the end-result. Moreover, many campuses are now moving towards integrated information services (for financial and library services amongst others) and it is important that any departmental endeavours be considered in this light. Academics may also profit from commercial administration aids such as the IBM PROFS mainframe system which provides facilities for handling electronic mail, keeping diaries and scheduling meetings.

9. SUMMARY

As we have discussed, computers are changing the ways of teaching microbiology and the use of computers is likely to increase. Instructors must make critical decisions regarding the best use of their own and student time. They must select the appropriate teaching aids, including CAI packages. CAI is inefficient for text presentation but is very good for presenting graphics and animation. Laboratory simulations and problem solving exercises are best suited for CAI because of the interactive potential of computer programs. In his use of computers it is important that the instructor should bear in mind the distinction between information transfer and the imparting of knowledge and understanding. Informed reviews of the history of the subject put a realistic perspective on the frankly limited contribution made by computers to education thus far (25).

Nevertheless, the breadth of opportunity, the doubts and the initial effort required to begin using computers in teaching should not obscure the fact that much useful work remains to be done and that much of this will be achieved by small, incremental advances. Computers form part of a revolution in educational technology whose impact on students cannot be gauged at the present time but is likely to be substantial.

10. REFERENCES

1. Woodhouse,D. and McDougall,A. (1986) *Computers: Promise and Challenge in Education*. Blackwell Scientific Publications, Melbourne.
2. Robertson,L.A. and Luyben,K.Ch.A.M. (1987) *TIBTech.*, **5**, 125.
3. Gardner,N. (1988) *Computer Educ.*, **12**, 23.
4. Roszak,T. (1986) *The Cult of Information*. Pantheon Books, New York.
5. Cooke,D., Craven,A.H. and Clarke,G.M. (1985) *Statistical Computing in Pascal.* Edward Arnold, London.
6. Papert,S. (1980) *Mindstorms: Children, Computers and Powerful Ideas*. Basic Books, New York.
7. Crabbe,M.J.C. (1985) *Methods Biochem. Anal.*, **31**, 417.
8. Bungay,H.R. (1971) *Proc. Biochem.*, **6**, 38.
9. Bazin,M.J. and Saunders,P.T. (1984) *Binary,* **1**, 30.
10. Bushell,M. (1987) *Binary,* **10**, 22.

11. Barber,P. (1987) *Author Languages for CAL.* Macmillan Education, Basingstoke.
12. Bertino,E. (1985) *Interfaces in Computing,* **3**, 37.
13. Rambally,G.K. and Rambally,R.S. (1987) *Computer Educ.*, **11**, 149.
14. Seiden,P. (1987) *Academic Computing,* Spring 1987, 10.
15. Stewart,D.J. (1985) *Binary,* **4**, 30.
16. Stewart,D.J. (1986) *Binary,* **7**, 29.
17. Cooper,N.S. (1987) *Binary,* **12**, 35.
18. Schaffer,A.G. (1986) *Binary,* **9**, 24.
19. Whiting,J. (1985) *Computer Educ.*, **9**, 154.
20. Whiting,J. (1986) *Computer Educ.*, **10**, 281.
21. Wimpenny,J.W.T. (1986) *Binary,* **7**, 22.
22. Johnston,V.M. (1987) *J. Comput.-Assisted Learning,* **3**, 40.
23. Self,J. (1985) *Microcomputers in Education: A Critical Appraisal of Education Software.* Harvester Press, Brighton.
24. Ross,P. (1987) *J. Comput.-Assisted Learning,* **3**, 194.
25. O'Shea,T. and Self,J. (1983) *Learning and Teaching with Computers.* Harvester Press, Brighton.

11. ADDENDUM

Future developments in the UK in the use of computers in teaching microbiology and related disciplines will be promoted by a national coordinator appointed by the Computer Board. Part of this person's remit will be to maintain a bulletin board and courseware catalogue on NISS (National Information for Software and Services) which is accessed via JANET (Joint Academic Network). For further information contact one of the authors, PGGM. NISS also carries files from CHEST (Combined Higher Education Software Team) that describe current software discounts available to the academic community.

CHAPTER 6

Computer simulation and mathematical models

J.I.PROSSER

1. INTRODUCTION

A mathematical model is a mathematical representation of a particular process and usually consists of an equation, or set of equations, describing that process. The use of such models in the physical, chemical and engineering sciences is commonplace. For example, the gas laws, Ohm's law, the basic laws of mechanics are all simple mathematical models, although the term is seldom used in such contexts. Many aspects of microbiology, and biology in general, are studied only qualitatively. This results in part from inherent biological variation which prevents, or at least makes difficult, formulation of precise relationships observed in physical phenomena. More importantly, biologists traditionally tend to have a non-mathematical background inducing, in some cases, a resistance to quantitative theoretical approaches to biological problems. The description of biological systems in terms of mathematical equations has consequently been sufficiently rare as to merit the term mathematical model.

Simulation of a system is the use of such a model to describe the behaviour of a system with respect to time and involves a number of steps. First, both the system to be modelled and the purpose of the study must be clearly defined. A suitable model is then constructed and predictions of the model are generated usually, but not necessarily, using a computer program over a series of time intervals. Tests must then be performed to determine that the program is correct and finally results of the simulation runs are analysed with respect to the questions posed at the onset of the study. Simulation, therefore, involves the study of dynamic rather than static models. The calculation of the specific growth rate of a population at a particular substrate concentration using the Monod equation (described below) would not constitute a simulation. The incorporation of this equation in a model describing the rate of change of population size with time as substrate was utilized would be considered a simulation.

2. DESCRIPTIONS AND FUNCTIONS OF MATHEMATICAL MODELS

Models of microbiological systems may be classified in terms of a number of contrasting categories. The most important distinction is between mechanistic and empirical models, which highlights an important difference in the function of a model. Mechanistic models describe a system in terms of the fundamental mechanisms which are believed to control that system. Empirical models merely describe the behaviour of the system on the basis of experimental data. The latter involve some form of curve fitting or regression analysis and the use of empirical models outside the range of experimental data on which they are based is dangerous. They do not in themselves increase our knowledge of the biology of the system, but may guide further studies. An empirical model, therefore, is an end in itself and has a descriptive function only. A mechanistic model represents a theory regarding the way in which a particular system works and mathematical representation increases the precision with which the theory is stated. Although the assumptions implicit in the theory, and the model, may be derived from experimental observations, further experimentation is required to test predictions of the model. The validity of the model, and hence of the assumptions, may then be assessed on the basis of the quality of fit between experimental and predicted results. The mechanistic model therefore plays a central role in a scientific investigation and in general would be more complex than an empirical model, which is likely to be of greater importance for technical or routine predictive use. This difference may be illustrated with reference to two basic equations describing microbial growth. The Monod equation relates the specific growth rate (μ) of a population to the concentration (s) of a limiting substrate and has the form

$$\mu = \frac{\mu_m s}{Ks + s} \tag{1}$$

Where μ_m and Ks are the maximum specific growth rate and saturation constant for growth respectively. The rate of change of biomass concentration (x) and substrate concentration with time (t) may be represented by the following two differential equations

$$\frac{dx}{dt} = \frac{\mu_m s x}{Ks + s} \tag{2}$$

$$\frac{ds}{dt} = \frac{\mu_m s x}{Y(Ks + s)} \tag{3}$$

Yield coefficient (Y) represents the amount of biomass formed per unit mass of substrate used. This model is mechanistic in that it considers biomass formation to result from substrate utilization and describes specific growth rate and substrate utilization in terms of biological constants (μ_m, Ks, Y) which may be determined experimentally. The mechanistic basis may be extended further if one assumes that uptake or utilization of the substrate is limited by the rate of a single enzyme reaction, in which case the dependence of μ on s will be analogous to that of the velocity of an enzyme reaction on s as described by the Michaelis–Menten equation.

An empirical model for growth is the logistic equation which has the form:

$$dx/dt = \mu_m x\,(1 - kx) \tag{4}$$

This model assumes, arbitrarily that specific growth rate decreases as a linear function of biomass concentration until a limiting value, $1/k$, is reached. It describes the growth of many microorganisms in batch culture well but the constant k has no biological meaning. A good or a bad fit between experimental and predicted results does not increase our knowledge of the system, but the former may lead us to investigate the relationship further. For example, a mechanistic model based on product inhibition (1) does lead to a relationship equivalent to Equation 4. A good fit with Equations 2 and

3, however, indicates that assumptions implicit in the Monod equation are correct; a single substrate and even a single enzyme may be limiting growth. Under certain conditions the Monod equation does describe microbial growth well. In others there are discrepancies, but even a poor fit gives us information by indicating situations where our assumptions may not be correct. A mechanistic model is therefore not a final statement but a theory which may require continual modification and refinement as more critical tests are performed.

Further distinctions arise from the particular requirements of the system under consideration. Equations 2–4 are unstructured, distributed models in which biomass is considered to have constant, uniform composition and to be averaged throughout the system. A structured model might consider biomass to be compartmentalized between DNA, RNA, protein and other cell components while a segregated model would describe the concentration of individual cells rather than of biomass. The models described above are also deterministic in that for a given set of inputs (initial conditions and growth constants) the behaviour of the system is determined solely by the differential equations. This contrasts with stochastic models in which the system behaviour varies randomly. For example, rather than each cell doubling at a fixed time, t_d, after birth, there will be a distribution of doubling times around t_d, with some cells having shorter and some longer doubling times.

The classification of mathematical models and examples of the different types are described by Roels and Kossen (2) and in the following sections. Distinction between the different categories is often blurred. Certain aspects of the system may be modelled satisfactorily by a mechanistic model while others may require empirical modelling, through lack of information on relevant mechanisms. Similarly, stochastic models may be constructed by introducing randomness into components of deterministic models. The majority of mathematical models of microbiological systems are mechanistic, deterministic and distributed and consist of ordinary or partial differential equations. Simulation of such models will form the basis for this chapter.

3. APPLICATION OF MATHEMATICAL MODELS TO MICROBIOLOGICAL PROCESSES

Mathematical modelling has now been introduced to most areas of microbiology and a complete review is outside the scope of this chapter. A selection of major areas of application will be provided to illustrate the scope and potential of the technique.

The major initial impact was in the field of microbial population growth and this has provided the basis for more applied aspects of modelling. Initially growth was described by unstructured models, with growth rate functions such as the Monod equation, but deviations from behaviour predicted by these models has led to the introduction of more complex, often structured models. The relationship between growth of individual cells and that of populations is described by models of microbial age and size distributions, which are often stochastic in nature (3). In addition, attempts have been made to describe the complex interactions between biochemical reactions occurring within a cell (4). Basic growth models must also be modified to consider specific features of particular organisms. For example, the mycelial growth form of filamentous fungi and actinomycetes requires special consideration as does the utilization of light, rather than a chemical energy source, by algae (5,6).

Basic growth models have now been extended to incorporate the effects of environmental parameters such as temperature and pH and features such as motility (6). These models are useful because they can further our understanding of microbial physiology and biochemistry but in addition they fulfil a predictive role as, for example, in the food industry where the effect of low temperature on growth is an important factor in controlling food spoilage.

Ecological applications range from the investigation of interactions between two microorganisms to the complexities of nutrient cycling in natural environments. A simple example of the former is given below for competition between two organisms for a growth limiting substrate in a chemostat. Quantitative studies of microbial interactions in which simple laboratory ecosystems have been used to test model predictions have greatly increased our understanding of the interactions between microorganisms in natural environments (7). The cycling of nutrients in aquatic and terrestrial environments poses enormous problems both for experimentalists and theoreticians, but it is impossible to consider quantitatively the many complex interactions between microorganisms and with their environment without some form of modelling. A systems analysis approach is usually adopted in which the system is divided into sub-systems: the latter can then be modelled in isolation before recombining to form a general process description. The problem with many such models is their complexity and the consequent number of biological constants which makes critical experimental testing difficult. Interactions with the physico-chemical environment (ion exchange, surface growth, leaching, light penetration, etc.) all combine to make these models more complex still.

Computer simulation is now routinely used for industrial fermentations and offers several advantages. Mechanistic models increase our understanding of the physiology of microbial growth, the kinetics of product formation and the factors exerting the greatest control of these processes. Modelling may be used to assess the economic costs of a process, in terms of raw materials, fermenter costs, fermenter design and energy costs. It also has an important role in the scale-up of laboratory fermentations, determination of optimal fermentation conditions for biomass and product formation and computerized control. Until recently, the latter referred to control of physical parameters such as temperature but the development of reliable monitoring systems for substrate, product and biomass concentrations offers great potential in the control of fermentation processes with on-line simulation and parameter estimation leading to optimal efficiency. These features are comprehensively reviewed by Kleinstreuer and Poweigha (8), Votruba (9), Harder and Roels (10) and in Chapter 8.

Most of these models consist of differential equations whose mathematics have been studied extensively because of their use in describing physical phenomena. Many biological phenomena, however, do not involve continuous changes and special techniques are required in such cases. One such technique is catastrophe theory (11) which has been used to model the sudden change in predator-specific growth rate which occurs in prey–predator interactions. This requires use of completely different mathematical techniques and will not be discussed here.

4. SIMULATION

Simulation involves the provision of values for initial conditions and constants within the model equations and calculation of changes in system components with respect to time. Simulation of bacterial growth would therefore involve determination of biomass concentration as a function of time. During exponential growth in batch culture, the rate of change of biomass concentration depends on the specific growth rate, μ (assumed constant), and the biomass concentration:

$$dx/dt = \mu x \quad (5)$$

Integration leads to the equation:

$$x = x_0 e^{\mu t} \quad (6)$$

where x_0 is the initial biomass concentration. Simulation of this simple model merely requires the calculation of x for a series of values of t after substitution of values for x_0 and μ. In practice the differential equations describing biological systems are usually more complex than Equation 5 and analytical solution, by integration, is impossible. This has two important consequences. First, it prevents immediate identification of specific relationships between variables and growth constants, for example from Equation 6 it is obvious that a plot of ln x versus t will be linear with slope μ and intercept x_0. Second, simulation of the model by simple substitution of values into the algebraic equation is impossible and numerical integration must be used to generate predictions. Numerical approximation methods necessarily involve errors and, although these can be controlled and minimized, they may not be reliable in all situations. They also require a large number of iterative arithmetic calculations and it was only the advent of powerful digital computers in the 1960s and 1970s which allowed their widespread application. It is now possible to carry out simulations with reasonable speed and efficiency on a microcomputer. It must be emphasized, however, that although numerical methods are often necessary they should be used only as a last resort. Analytical solutions are always preferable and provide important insights into the behaviour of the system even when this requires relaxing assumptions used in initial model building. Application of even simple mathematical techniques may also save time. Microbial growth in a chemostat may be modelled by Equations 2 and 3 with additional terms for washout of biomass and substrate and for input of fresh substrate. Analytical solution of this set of equations is not possible and changes in x and s may only be determined using numerical methods. However, on establishment of a steady-state, x and s are constant, dx/dt and ds/dt may be set equal to 0 and algebraic expressions may then be obtained relating steady-state biomass and substrate concentrations to dilution rate. These relationships may have been obtained by repeated simulation of the model at different dilution rates but this would have been extremely inefficient and time consuming in comparison with steady-state analysis.

5. ANALOGUE COMPUTERS

Before the advent of digital computers, differential equations were solved using analogue computers in which the values of system variables are represented by continuously variable voltages in electronic circuits. Each mathematical operation—summation, multiplication, integration, etc.—is carried out by a component within the circuit. Simulation of a model requires wiring of these components and setting voltages representing initial conditions and constants. This is a complicated process which, along with the expense and limited accuracy of analogue computers, has led to its replacement using digital computers. Analogue computers do possess an advantage in fast computation and greater on-line control, however, and new integrated circuits have improved instrumentation and control. Occasionally a case for combining analogue and digital computers can be made by using the fast simulation computations of the former and the storage and data output facilities of the latter.

6. NUMERICAL APPROXIMATION METHODS FOR ORDINARY DIFFERENTIAL EQUATIONS

6.1 **Theoretical aspects**

Numerical analysis is concerned with the development of approximation techniques for the solution of mathematical problems and with assessment of the value of these techniques. It depends on the ability of digital computers to carry out arithmetical operations at high speed and low cost and, to a lesser extent, on their ability to store and retrieve information quickly. Practical numerical analysis involves the development of computer algorithms to carry out approximation techniques; their value is assessed in terms of accuracy, stability, speed and hence cost.

A detailed consideration of numerical techniques can be found in standard texts (12,13). The aim here is to describe the basis of these techniques, to demonstrate computer algorithms used for their practical solution and, most importantly, to highlight sources of error and dangers in using approximation techniques. Appreciation of the latter is necessary to enable efficient checking of simulation programs, efficient programming in terms of time and cost and to prevent the generation of false predictions.

Solution of ordinary differential equations requires the calculation of values of a dependent variable for certain values of an independent variable. The solution may be of the boundary-value type in which boundary conditions are distributed between two or more points. Microbiological problems are usually of the initial-value type where one provides an initial value for the dependent variable and determines its change up to a certain value of the independent variable.

The techniques will be illustrated by considering the rate of change of a biological variable, x (e.g., biomass concentration), with time, t. This may be represented by a first order differential equation of the type:

$$\mathrm{d}x/\mathrm{d}t = f(t,x) \tag{7}$$

Equation 7 states that x is some function (f) of both x and t. An example of such an equation is the logistic equation, Equation 4. The techniques to be described are capable of solving first-order differential equations of this type. Equations of a higher order are encountered less frequently but may be solved using the same techniques following reduction to first-order form (which is almost always possible by introduction of new variables).

6.1.1 *Euler method*

The basic procedure for numerical solution of a differential equation such as Equation 7 involves calculating changes in x with respect to small changes in t. The simplest technique, Euler's method, calculates the change in x as the product of the function $f(t,x)$ and the time interval or step size, h:

$$x_{(\mathrm{t}+\mathrm{h})} = x_{(\mathrm{t})} + hf(t,x) \tag{8}$$

where $x_{(t)}$ and $x_{(t+h)}$ are values of x at times t and $(t + h)$. Subsequent values of x may be calculated iteratively using Equation 8:

$$x_{(t+2h)} = x_{(t+h)} + hf(t + h, x_{(t+h)})$$

$$x_{(t+3h)} = x_{(t+2h)} + hf(t + 2h, x_{(t+2h)})$$

Equation 8 may be simplified by using subscript i to represent the number of iterations or steps:

$$x_{i+1} = x_i + hf(t_i, x_i) \qquad i = 0,1,2.......... \tag{9}$$

This equation can now be used to solve the logistic equation, with an initial value of $x = 1$, $\mu_m = 0.5$, $k = 0.01$ and step size, $h = 0.5$:

$$x_1 = 1 + 0.5 * [0.5 * 1 (1 - 0.01 * 1)] = 1.2475$$

$$x_2 = 1.2475 + 0.5 * [0.5 * 1.2475 (1 - 0.01 * 1.2475] = 1.55548436$$

A computer program for iteration of this process is provided in Section 6.2.

Euler's method calculates values of x over 1 step, that is, values of x_{i+1} are calculated from a knowledge of values of x_i. The accuracy of the technique increases with decreasing values of h, but this necessarily involves increased computation time and increased cumulative errors (see below). Accuracy can be improved by increasing the rate at which errors decrease as h decreases and this requires use of multistep methods. These calculate x_{i+1} from a knowledge not only of x_i but also of x_{i-1}, x_{i-2}, etc. the most common are two-step methods which use values of x_i and x_{i-1} to calculate x_{i+1}. Equation 9 is then replaced by:

$$x_{i+1} = x_{i-1} + 2hf(t_i,x_i) \qquad i = 1,2,3....$$

A disadvantage of multistep methods is their requirement for an initial value for x_{i-1}, which may be calculated using a Taylor series expansion or by the Runge–Kutta method described below. The technique is said to be non-self-starting because of this requirement and for this and other reasons Runge–Kutta techniques are more commonly used.

6.1.2 *Runge–Kutta method*

Runge–Kutta (R–K) type method are one-step techniques, they do not require values from two or more preceding time steps, they are therefore self-starting and can also achieve greater accuracy. They replace Equation 9 with an equation of the form:

$$x_{i+1} = x_i + \alpha_0 K_0 + \alpha_1 K_1 + ...\alpha_n K_n \tag{10}$$

where $K_0 = hf(t_i\ x_i)$.

$$K_1 = hf(t_i + \mu_1 h, x_i + \lambda_{10} K_0)$$

$$K_2 = hf(t_i + \mu_2 h, x_i + \lambda_{20} K_0 + \lambda_{21} K_1)$$

$$K_n = hf(t_i + \mu_n h_1, x_i + \lambda_{n0} K_0 + ... \lambda_{n,n-1}, K_{n-1})$$

This reduces to Euler's Equation 9 if $\alpha_0 = 1, \alpha_1, \alpha_2, \alpha_3...... = 0$. The rationale behind this technique is that from a knowledge of values of t_i and x_i we can calculate all the values of x in the interval x_i to x_{i+1}. Whereas Euler's method chooses just one pair of values, t_{i+1}, x_{i+1}, R–K methods improve accuracy by incorporating additional information on other pairs of values. The values of the coefficients α, μ and λ are chosen to maximize accuracy. Accuracy can also be increased by increasing the number of terms in Equation 10 (i.e. increasing the order of the integration routine) with obvious increased cost in terms of computational time. The most common method involves four K terms ($K_0 - K_3$), and this is therefore called a fourth order Runge–Kutta method. The most convenient values for coefficients lead to the following version of Equation 10:

$$x_{i+1} = x_i\ (K_0 + 2K_1 + 2K_2 + K_3)/6 \tag{11}$$

where
$$K_0 = hf(t_i, x_i)$$
$$K_1 = hf(t_i + h/2,\ x_i + K_0/2)$$
$$K_2 = hf(t_i + h/2,\ x_i + K_1/2)$$
$$K_3 + hf(t_i + h,\ x_i + K_2)$$

This method is therefore self-starting, has good accuracy and may be applied to the simultaneous solution of a set of ordinary differential equations. It has the properties required for most biological applications and is the technique of choice. Care must, however, be adopted in application of this technique, as with all others, with regard to errors and stability.

6.1.3 *Error generation*

There are two major sources of error in numerical approximation techniques. Firstly, truncation error arises from the limitations of the approximation technique used. For example, Equation 9 contains only the leading two terms of a series of expansion in h^2, h^3.........h^n and ignores all terms of order h^2 and above. Valid approximations possess the property that an exact solution would be obtained if sufficiently high values of n were used. In ignoring terms containing h of order 2 and above we are introducing a truncation error of order h^2 (if h is small). This error will occur regardless of the accuracy with which the arithmetical calculations are carried out at each step. Truncation errors for multistep methods are of order at least as small as h^3. The errors generated may be local or global. Local errors are those which occur at each step while global or total errors are cumulative over the whole period of integration. Local errors are usually reduced by reducing h and, for R–K methods, by increasing the order of the integration method, that is by increasing the number of K terms in Equation 10. Global errors may not be reduced by reducing h and in fact in some situations may actually increase.

Ultimately the arithmetical operations involved in approximation techniques are carried out with a certain precision, defined by the number of significant figures used in addition, subtraction, multiplication, etc. The second source of error, round-off, results from rounding off the last significant figure during iterative computations. These errors are cumulative over a series of steps and increase with the order of the integration rule. They also increase with decreasing step size, as this results in a greater number of steps, with each step requiring greater precision. The factors that decrease round-off error

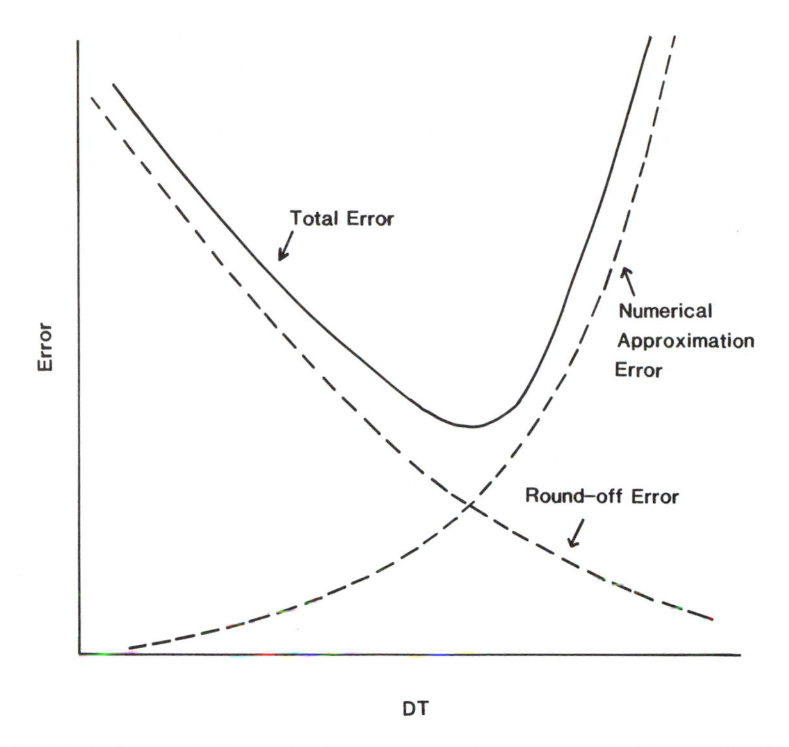

Figure 1. Variation in numerical approximation error (truncation error), round-off error and total error with step size, DT. (Redrawn from ref. 14).

are unfortunately those which increase truncation error as illustrated in *Figure 1*. There usually exists, therefore, a step size at which the combined error is minimized. The effects of step size for the different methods will be illustrated below in practical examples and should always be investigated in simulation studies.

6.1.4 *Stability*

A further important feature of approximation techniques is their stability. Instability occurs when a series of approximation calculations leads to overflow. A particularly common situation in simulating microbial growth and substrate conversion occurs when substrate concentration decreases towards zero, as a result of growth, and then becomes negative, even if specific growth rate is a function of substrate concentration. This may occur when the system response time is smaller than the integration step size. Thus with an exponentially increasing biomass concentration utilizing a decreasing and small substrate concentration, large transients exist and the substrate concentration may 'overshoot' the zero value, becoming negative, rather than approaching zero asymptotically. Accurate numerical solution of the equations under these conditions requires an increasingly small step size and therefore more computational time. Instability may also result from round-off errors, which cannot be reduced by reducing step size. Further types of instability are discussed in Hildebrand (12) and other standard texts, some arising from effects at particular values of x and others from cumulative or step size phenomena.

A specific, and not uncommon, problem is that of stiff systems in which instability arises from the existence of sub-systems with a wide range of time constants. This occurs in chemical reaction systems with a wide range of values for rate constants or for concentrations of interacting chemical species. Special techniques are required for stiff systems and are usually provided for in simulation software.

6.1.5 *Variable step size*

Numerical approximation involves a balance between accuracy (integration routine, order number and step size) and computational speed and cost. When integrating an equation over a series of time intervals, the rate of change of system components will vary and the step size may also be varied with no loss of accuracy. This is achieved most simply by setting a minimum and maximum on some form of error measure and doubling or halving the step size if the error falls outside this range at any particular step. Doubling the step size presents no problem, but a halving requires provision of intermediate values calculated by a self-start procedure. Modifications of the R−K method, the Runge−Kutta−Merson and Runge−Kutta−Fehlberg methods, estimate the local per step error by subtracting solutions and second- and fourth-order R−K methods at each time step. Not only do variable step size routines save computational time, they also reduce chances of overflow resulting from small system response times, by automatically reducing step size as transients increase.

6.2 **Practical aspects**

In this section a description will be given of computer programs demonstrating use of the Euler and R−K methods for solution of ordinary differential equations. The programs have been written in BASIC for a BBC Model B+ Microcomputer and solve the logistic equation, Equation 4, demonstrating how the approximation equations are converted into computer algorithms and illustrating the effects of step size on errors.

6.2.1 *Euler method*

The logistic equation is presented again below along with its analytical solution:

$$\mathrm{d}x/\mathrm{d}t = \mu_{\mathrm{m}}{}^{x(1 - kx)}$$

$$x = x_0/[(1 - kx_0)\exp(-\mu_{\mathrm{m}}t) + kx_0] \tag{12}$$

where x_0 is the initial biomass concentration. *Program 1* simulates the model using the analytical solution with initial values $x_0 = 1.0$, $\mu_{\mathrm{m}} = 0.5$/h, $k = 0.01$, at time intervals (TINT) = 1.0 h up to TMAX = 25.0 h. (All programs are given in the Appendix to this chapter.) These values are set in the procedure INIT, lines 150−230. This initialization procedure places the values for each time interval in an array T and also sets values for $t = 0$ and step size for integration H = 0.5 h, for use in approximation routines. The analytical simulation is carried out in procedure ANALYT, lines 260−310. Values for x at each time interval are placed into array X1.

Simulation by the Euler method is carried out by procedure EULER, lines 340−420. Equation 8 is equivalent to the first statement on line 380, where $f(t,x) = \mu_{\mathrm{m}}x\,(1 - kx)$. Time, T, is incremented by the step size H at each iteration and values of x are placed in array X2 at time intervals contained in array T. A comparison of analytical and approximate results are given for two step sizes, 0.5 h and 0.1 h, in *Table 1*.

6.2.2 *Runge–Kutta method*

Simulation by the R–K method is carried out in procedure RUNGE (lines 450–570) which itself calls a procedure RK (lines 600–850). These procedures use arrays Y, F, S and P which must be declared (line 30) with dimension N, where N = the number of differential equations, in this case = 1. The aim is to calculate Equation 11 iteratively, with step size H, from T = 0 to T = TMAX. The function $f(t,x)$ is calculated in line 520 and stored in array F. The array Y contains the value of x_i at the beginning of the step and x_{i+1} at the end of the step. Procedure RUNGE makes five calls to procedure RK. The first precedes calculation of $f(t_i,x_i)$. The four subsequent calls calculate K_0, K_1, K_2, K_3 and finally x_{i+1}. During these calls T is incremented by the step size H in two stages. Values for x are stored in array X3 at the required time intervals: results are printed out by procedure OUTPUT (lines 880–990) and are given in *Table 1* for two step sizes 0.5 h and 0.1 h.

6.2.3 *Comparison of methods*

Table 1 provides an estimate of the approximation error, calculated as the difference between approximate and analytical solution values divided by the latter and expressed as a percentage (procedure OUTPUT). With a step size of 0.5 h the Euler method generates errors of more than 20%, while the R–K method has an accuracy approximately 3–4 orders of magnitude better. The accuracy of both methods varies throughout the simulation and is greatest when the changes between time intervals are smallest, that is when transients are smallest. A step size of 0.1 h decreased the percentage error for the Euler and R–K methods by approximately 1 and 3 orders of magnitude respectively and led to errors of no greater than 5.5% for the Euler method. At this small step size the accuracy of the R–K method is likely to be limited by round-off errors due to the precision of the computer rather than by truncation errors. To test the speed of the two methods, simulations were run with a step size of 0.01 h. The program took 43 and 340 sec for the Euler and R–K routines, respectively.

Increased computational time for the R–K technique is therefore more than compensated for by greater accuracy at each time step and greater stability. The required accuracy can be obtained with larger step sizes, reducing the overall computational time significantly and R–K techniques are routinely used for simulation of ordinary differential equations in microbiological mathematical models.

6.3 **Simulation of competition in a chemostat**

The system to be considered is a chemostat containing two populations of bacteria, biomass concentrations x_1 and x_2, competing for a single limiting nutrient, concentration s. Fresh medium is supplied at a concentration s_r and the dilution rate is D. The situation may be conveniently modelled by three differential equations describing rates of changes of x_1, x_2, and s

$$\mathrm{d}x_1/\mathrm{d}t = \mu_1 x_1 - \mathrm{D}x_1 \tag{13}$$

$$\mathrm{d}x_2/\mathrm{d}t = \mu_2 x_2 - \mathrm{D}x_2 \tag{14}$$

$$\mathrm{d}s/\mathrm{d}t = \mathrm{D}s_r - \mu_1 x_1/Y_1 - \mu_2 x_2/Y_2 - \mathrm{D}s \tag{15}$$

Specific growth rates, μ_1 and μ_2, are described by the Monod equation (Equation 1) and Y_1 and Y_2 represent yield coefficients for the two populations. Simulation of this model requires three steps:

(i) input of values for constants, initial conditions and experimental control parameters (D and s_r);
(ii) solution of the differential equations;
(iii) output of results.

The following values will be provided for step 1:

$\mu_{m1} = 0.5/h$ $\qquad \mu_{m2} = 0.25/h$

$Ks_1 = 5.0\ \mu g/ml$ $\qquad Ks_2 = 1.0\ \mu g/ml$

$Y_1 = 0.4\ g/g$ $\qquad Y_2 = 0.6\ g/g$

$D = 0.2/h$ $\qquad s_r = 50\ \mu g/ml$

$x_{10} = 1\ \mu g/ml$ $\qquad x_{20} = 1.0\ \mu g/ml$

$s_0 = 0.0\ \mu g/ml$

Subscript 0 refers to values at time = 0.

Table 1. Results of simulation of the logistic equation by three methods (*Program 1*).

Time	*Biomass concentration*			*% Error*	
	Analytical	*Euler*	*Runge–Kutta*	*Euler*	*Runge–Kutta*
a					
Step size = 0.5					
0	1	1	1	0	0
1	1.6380946	1.55548436	1.63807455	5.04306912	1.22392484E-3
2	2.6723631	2.41349069	2.67229961	9.68702209	2.37587419E-3
3	4.33090058	3.73034175	4.33075288	13.8668347	3.41029728E-3
4	6.94531596	5.73162404	6.94501998	17.4749706	4.26162941E-3
5	10.9572051	8.72760057	10.956674	20.3482964	4.84720563E-3
6	16.8664789	13.1115952	16.8656212	22.2624043	5.08507788E-3
7	25.0655404	19.3128556	25.0643046	22.9505715	4.93026674E-3
8	35.5460987	27.6641562	35.5445292	22.1738609	4.41533603E-3
9	47.6237951	38.1658433	47.6220512	19.8597188	3.66169614E-3
10	59.9859602	50.2276861	59.9842557	16.2675967	2.84139741E-3
11	71.1951342	62.6226596	71.1936345	12.0408152	2.10635611E-3
12	80.2957153	73.8710786	80.2944871	8.00121983	1.5295754E-3
13	87.0442866	82.8877837	87.0433237	4.77515872	1.10619924E-3
14	91.7198683	89.3652193	91.7191357	2.56721804	7.98802165E-4
15	94.8087153	93.6353553	94.8081721	1.23760776	5.72981852E-4
16	96.7856705	96.2845255	96.7852772	0.517788432	4.06332306E-4
17	98.0254375	97.8642649	98.025159	0.164419158	2.84143088E-4
18	98.7929897	98.7835904	98.7927961	9.5141314E-3	1.95991324E-4
19	99.2644178	99.3108976	99.2642853	−4.6824192E-2	1.33483003E-4
20	99.5525518	99.6108187	99.5524623	−5.8528764E-2	8.98986239E-5
21	99.7281303	99.780588	99.7280705	−5.26006657E-2	5.99763185E-5
22	99.8349261	99.8764227	99.8348865	−4.15651583E-2	3.97026274E-5
23	99.8998125	99.9304376	99.8997865	−3.06557884E-2	2.6073352E-5
24	99.9392093	99.9608553	99.9391922	−2.16591896E-2	1.70572977E-5
25	99.9631197	99.9779761	99.9631087	−1.48618196E-2	1.10607408E-5

b

Step size = 0.1

0	1	1	1	0	0
1	1.6380946	1.61919062	1.63809456	1.15402257	2.30258767E-6
2	2.6723631	2.61213322	2.67236298	2.25380582	4.39111903E-6
3	4.33090058	4.18924387	4.3309003	3.27083722	6.32221479E-6
4	6.94531596	6.6563956	6.94531541	4.15993114	7.93834243E-6
5	10.9572051	10.4252653	10.9572042	4.85470341	8.94161728E-6
6	16.8664789	15.9777736	16.8664773	5.26906237	9.40903955E-6
7	25.0655404	23.7334728	25.0655381	5.31433802	9.12538972E-6
8	35.5460987	33.7899998	35.5460958	4.94034215	8.09068847E-6
9	47.6237951	45.6288139	47.6237919	4.18904276	6.66462496E-6
10	59.9859602	58.0571479	59.9859571	3.21543941	5.11726277E-6
11	71.1951342	69.6057902	71.1951315	2.23237719	3.80926501E-6
12	80.2957153	79.1633969	80.2957131	1.41018531	2.74656232E-6
13	87.0442866	86.3326021	87.0442849	0.817611991	1.95157252E-6
14	91.7198683	91.3186712	91.7198671	0.437415767	1.36469618E-6
15	94.8087153	94.6050947	94.8087145	0.214769931	8.80156454E-7
16	96.7856705	96.6947299	96.7856698	9.39607738E-2	6.46633708E-7
17	98.0254375	97.9930338	98.0254371	3.30563969E-2	4.56039623E-7
18	98.7929897	98.7880907	98.7929894	4.95881881E-3	3.01664344E-7
19	99.2644178	99.2706537	99.2644176	−6.28204756E-3	2.10162172E-7
20	99.5525518	99.5619644	99.5525517	−9.45492237E-3	8.98088151E-8
21	99.7281303	99.7372459	99.7281303	−9.14039691E-3	5.97671335E-8
22	99.8349261	99.8425046	99.8349261	−7.59099311E-3	5.97031992E-8
23	99.8998125	99.9056389	99.8998125	−5.83219716E-3	8.94966315E-8
24	99.9392093	99.94348	99.9392093	−4.27333019E-3	0
25	99.9631197	99.9661513	99.9631197	−3.03267029E-3	2.98133176E-8

Percentage error represents the percentage deviation from the analytical solution. Results are presented for step sizes (**a**) 0.5 h, (**b**) 0.1 h.

Three methods will be used to simulate this model. The first, based on the R−K routine in *Program 1,* was written in BASIC on a BBC Microcomputer. The second is written in FORTRAN and uses a NAG Library routine to solve the differential equations. The third uses a simulation language MIMIC, and will be preceded by a discussion of simulation languages.

Although the NAG library and MIMIC will be discussed in some detail they are merely representatives of a number of software libraries and simulation languages available on mainframes and, increasingly, on microcomputers. Information on such software, and on graphical packages which are useful for many aspects of simulation, can be obtained from the manufacturers' software catalogues and from software literature. Mainframes are generally supplied with a software for numerical methods, graphics and a continuous simulation language.

6.3.1 *Microcomputer simulation*

Program 2 is a modification of *Program 1* omitting procedures ANALYT and EULER. Procedure INIT is used to input values for growth constants, initial conditions and experimental control parameters. The differential equations are again solved using procedures RUNGE and RK. The arrays Y, F, S and P have now been dimensioned

Table 2. Results of simulation of model for competition using a Runge–Kutta technique with a step size of 0.5 h from $t = 0$ to $t = 100$ h (*Program 2*).

Time	*X1*	*X2*	*S*
Step size = 0.5			
0	1	1	0
5	2.15556846	1.13669822	25.8554255
10	6.56102419	1.39649371	25.0670716
15	15.4336609	1.64886478	6.38582616
20	17.1785292	1.64382275	3.47450257
25	17.4594243	1.59121532	3.39058984
30	17.586352	1.53526499	3.36173483
35	17.6565386	1.47950978	3.35100903
40	17.705039	1.42512853	3.34681282
45	17.7447403	1.37249147	3.34500706
50	17.7804091	1.32168556	3.34408712
55	17.8138269	1.27269912	3.34350213
60	17.8456762	1.22548696	3.34304969
65	17.8762343	1.17999336	3.3426551
70	17.9056284	1.13616012	3.34229059
75	17.9339298	1.09392947	3.34194559
80	17.9611883	1.05324497	3.34161589
85	17.9874448	1.01405187	3.34129964
90	18.0127366	0.976297134	3.34099584
95	18.0370987	0.939929499	3.34070382
100	18.0605645	0.904899389	3.34042307

Biomass concentrations, X1 and X2, and substrate concentration, S, are in μg/ml and growth constants and initial conditions are defined in the text.

to size three, with elements 1, 2 and 3 of array F evaluating functions for Equations 13, 14 and 15 with values of x_1, x_2 and s placed in array Y at each time interval. Predicted values are stored at the required time intervals in arrays X1, X2 and X3. Results are printed out by procedure OUTPUT and are presented in *Table 2*.

6.3.2 *Mainframe simulation using NAG library routine*

Program 3 is a FORTRAN program which uses a routine from the NAG (Numerical Algorithms Group) Library (15) to solve the differential equations. This is a typical example using software libraries written for numerical methods and statistical applications. Libraries of software routines were originally implemented on mainframe digital computers, but with development of microcomputer technology, subsets of these are now available for some machines. The NAG Library contains routines for solving ordinary differential equations for initial-value and boundary-value problems. The former are solved by several methods, including R–K methods, and the routine chosen, DO2BBF uses a Runge–Kutta–Merson variable step size method. Other routines contain higher accuracy and methods for stiff equations.

The routine is called by the main section of *Program 3* as a subroutine DO2BBF with the following arguments. T, TMAX, N and array Y have the same functions as in *Program 2*. TOL represents a tolerance for controlling the errors in approximation and controls the step size for integration. The chosen value of 0.00001 will give accuracy

Table 3. Results of simulation of the competition model using a NAG Library routine (*Program 3*).

Time	*X1*	*X2*	*S*
0	1.0000000	1.0000000	0.
5.0	2.1558512	1.1404863	25.8483828
10.0	6.5611882	1.4011301	25.0588592
15.0	15.4311805	1.6542894	6.3829593
20.0	17.1749357	1.6492114	3.4744960
25.0	17.4559253	1.5964366	3.3906294
30.0	17.5829707	1.5403073	3.3617741
35.0	17.6532822	1.4843732	3.3510371
40.0	17.7019345	1.4298180	3.3467548
45.0	17.7417439	1.3770110	3.3449641
50.0	17.7775880	1.3260428	3.3438770
55.0	17.8110010	1.2768949	3.3435735
60.0	17.8429442	1.2295296	3.3431417
65.0	17.8735810	1.1838880	3.3427971
70.0	17.9031042	1.1399132	3.3423461
75.0	17.9315190	1.0975457	3.3419456
80.0	17.9588821	1.0567291	3.3415744
85.0	17.9852161	1.0174080	3.3412779
90.0	18.0105696	0.9795296	3.3410260
95.0	18.0349612	0.9430422	3.3408597
100.0	18.0585464	0.9078983	3.3404700

Results are presented as in *Table 2*.

to approximately five decimal digits. IR determines whether the error is measured in terms of the number of correct decimal places (IR = 1), the number of correct significant digits (IR = 2) or a mixture of both (IR = 0). FCN is the subroutine which evaluates the functions from Equations 13–15 which are stored in array F. Subroutine FCN is provided by the user and is specific to the particular model being simulated. A second user-supplied subroutine, OUTPUT, is used to print out results at a time interval DT. The final two arguments, W and IFAIL, provide work space and information on errors detected by the subroutine. All real values and arrays must be defined as double precision.

Growth parameters and experimental control parameters are defined in DATA statements within subroutine FCN and all other values required by the subroutine and initial conditions are provided in the main program. Results of simulation of competition by *Program 3* are given in *Table 3* for a tolerance of 0.00001.

6.3.3 *Simulation languages*

Simulation languages are designed to facilitate input of data, solution of differential equations and output of results and are particularly useful for users with little or no previous programming experience. Both discrete and continuous simulation languages are available. The former consider changes at discrete intervals of time while continuous simulations describe the system behaviour as a continuous function from the beginning to the end of the simulation. Discrete simulation languages have a role in modelling stochastic processes. Most microbial applications involve continuous simulation which will be considered here.

Continuous simulation languages provide simplified statements for data input and output and integration, which are then translated into FORTRAN programs which are executed with results printed out and/or plotted. A wide range of languages is now available and at least one continuous simulation language is implemented on most mainframe computers. Basic structural features are common to all, and only one, MIMIC (16), will be described in detail. This has been chosen as one with which the author has experience and others are commonly used. For example CSMP (Continuous Simulation Modelling Program) is described in (17) and an example of its use in simulating penicillin fermentation is given in (18), while Clore (19) describes a powerful package (FACSIMILE) which combines solution of a set of ordinary differential equations with non-linear optimization of model parameters.

6.3.4 *Simulation using MIMIC*

Program 4 simulates the competition model using MIMIC. The first four statements are used to input initial conditions, growth constants and experimental control parameters. This is achieved using CON statements which contain parameters of the model which remain constant throughout the program. The first CON statement in this example defines UM1, UM2, KS1, KS2, Y1 and Y2 and instructs the program to read data values for these. Values must always be provided for TMAX and DT, the interval at which results will be printed or plotted. The PAR statement defines parameters which may vary between different simulations within the one program. In this example, the simulation is run for two dilution rates, D, using the same growth constants and initial conditions. Values defined by CON and PAR statements are printed with headings at the beginning of the results section. The differential equations are solved in the following five lines. Both algebraic and differential equations may be evaluated. Integral equations are preceded by INT, appear within brackets and are terminated by the initial value of the integrand. Termination of the simulation is controlled by a FIN statement and occurs when T = TMAX. Headings for columns of results are produced by a HDR statement and results are printed, at interval DT, by OUT statements. Results may also be plotted using PLO statements. In the example given, the first PLO statement plots X1 and X2 versus time and the second plots S versus time. Data values are placed at the end of the program and correspond to CON and PAR statements. The first line of data therefore provides growth constants in the first CON statement, those on the following line provide values for the second CON statement and so on. Parameter values are provided after those for constants.

Examples of numerical output are given in *Table 4*, and are seen to correspond closely with those in *Tables 2* and *3*. Plotted output is illustrated in *Figure 2*, and is explained in the figure legend.

6.4 **Choice of technique**

The three techniques described above give essentially similar results for the simulation of competition in a chemostat, although computational time taken for simulation and facilities available varies. The microcomputer simulation was presented to illustrate a computer algorithm for R−K methods described in Section 6.1.2 and to demonstrate the feasibility of its use on a microcomputer. Variable step size techniques could also

Table 4. Results of simulation of the competition model using MIMIC (*Program 4*).

Time	*X1*	*X2*	*S*
Step size = 0.5			
0	1	1	0
5	2.15556846	1.13669822	25.8554255
10	6.56102419	1.39649371	25.0670716
15	15.4336609	1.64886478	6.38582616
20	17.1785292	1.64382275	3.47450257
25	17.4594243	1.59121532	3.39058984
30	17.586352	1.53526499	3.36173483
35	17.6565386	1.47950978	3.35100903
40	17.705039	1.42512853	3.34681282
45	17.7447403	1.37249147	3.34500706
50	17.7804091	1.32168556	3.34408712
55	17.8138269	1.27269912	3.34350213
60	17.8456762	1.22548696	3.34304969
65	17.8762343	1.17999336	3.3426551
70	17.9056284	1.13616012	3.34229059
75	17.9339298	1.09392947	3.34194559
80	17.9611883	1.05324497	3.34161589
85	17.9874448	1.01405187	3.34129964
90	18.0127366	0.976297134	3.34099584
95	18.0370987	0.939929499	3.34070382
100	18.0605645	0.904899389	3.34042307

Results are presented as in *Table 2*.

be incorporated, thereby saving time and reducing the risk of overflow. Microcomputer simulation is sufficiently accurate but is still rather slow for mathematical models of the complexity usually encountered in microbiology. This situation is likely to change as faster machines become available and a microcomputer simulation package (Micromodeller, IRL Press, Oxford) is now available. Their use does relieve the user from dependence on a mainframe computer, whose response time may vary with user load. The implementation of software routines such as NAG will also increase the advantages of microcomputer simulations.

Library routines, such as those used in *Program 3* have significant advantages over those provided by the user, as in the simple example of *Program 2*. They allow assessment of error and subsequent variation in step size so that a tolerance may be specified. Failure diagnostics significantly aid in de-bugging which can be time consuming during program development. A range of routines is provided, designed to solve particular problems and using a variety of methods, including solution of stiff equations. All routines are written in standard format facilitating use of different techniques if necessary. Finally, and importantly, all routines are extensively tested before release and are therefore less likely to fail compared with those written, and entered, by the user. Checks on suitability must, however, be carried out as described below, as it is impossible to test fully any routine and errors may be generated which are specific to the simulation under investigation.

The major advantage of simulation languages is their simplicity, as can be seen by comparing listings of *Programs 3* and *4*, and the facilities offered in terms of printing

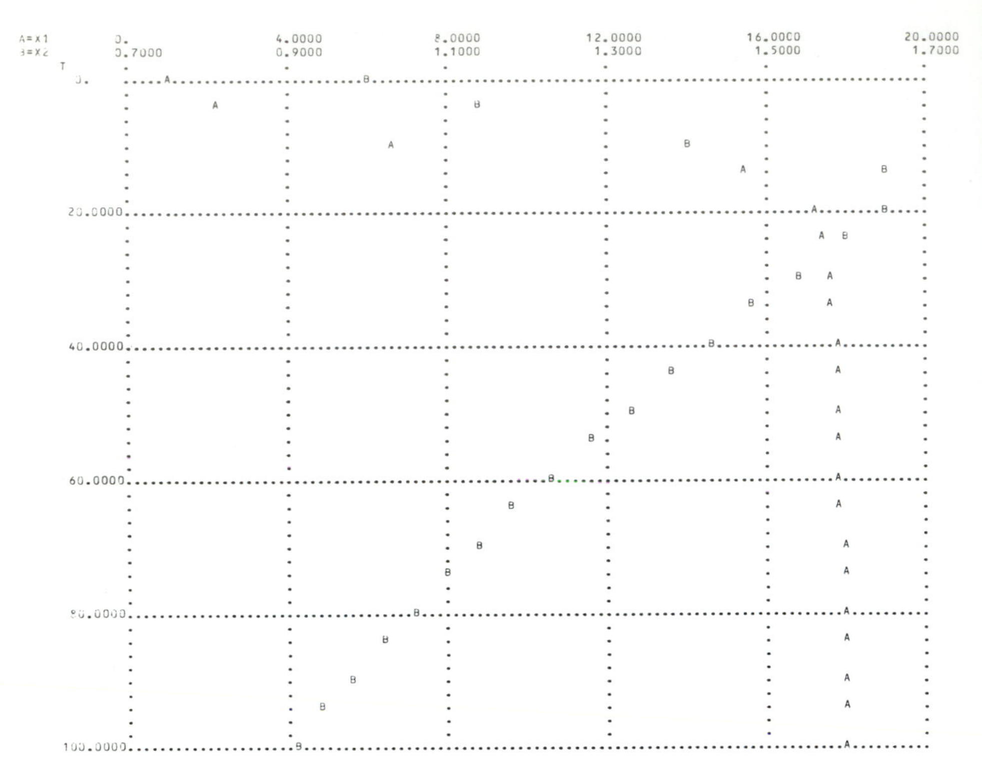

Figure 2. Example of plotted output from simulation of competition using the simulation language MIMIC. Biomass concentrations X1 and X2 (plotted as A and B) are plotted as a function of time.

and plotting of results. When available, they are excellent for the inexperienced user simulating straighforward models with no special requirements. The plotting facility is particularly useful and its reproduction in FORTRAN or BASIC programs would require considerable programming time unless ready made packages were available or one uses a graphics subroutine library. Additional features such as logical control variables and the ability to incorporate subroutines written in FORTRAN offer further advantages.

The disadvantages result from a lack of flexibility. Using MIMIC as an example, diagnostics for errors in programming MIMIC statements are good, but assessment of local and global approximations is not provided, nor of stability failures, and provision is not made for special situations such as stiff equations. Although FORTRAN subroutines may be incorporated, this increases program complexity and complications resulting from singularities or discontinuities present problems. These are best dealt with by users experienced in programming or by FORTRAN programs containing library routines in which discontinuities are considered by conditional statements: such discontinuities are frequently encountered in microbiological models.

The choice of technique therefore depends on the purpose of the study, the complexity of the model and the programming experience of the user. For experienced programmers, capable of writing plotting routines if necessary, the versatility of FORTRAN programs using mathematical and graphical software library routines probably provides the best course.

6.5 Program design

Sections 6.1.3 and 6.1.4 highlight the need for rigorous testing of the simulation technique and program before the model itself may be tested. The first step is to ensure that there are no programming errors and that the model is internally consistent. For example, if a substrate is being converted into a product, the total amount of product and remaining substrate should be equal to the amount of the substrate supplied. Mass balance calculations of this sort may be difficult in complex models consisting of a system of several differential equations, but must always be carried out. The significance of approximation errors should be investigated by varying the step size, or for library routines by varying the tolerance. Reduction in step size should lead to convergence of the solution and a decrease in global error up to the point where machine precision limits accuracy. Precision on microcomputers is of the order of nine significant figures while use of Double Precision on mainframe computers provides substantially greater accuracy, of the order of 18 significant figures. Overflow conditions should be corrected and statements inserted into the program, where possible, to trap unrealistic solutions. Step sizes can then be chosen which optimize the simulation in terms of computational time required but the user should always be wary of errors which may arise under specific sets of conditions.

Program 2 was intentionally written without sophisticated input and output facilities. Improvements which could be made include the output of initial values, growth constants and experimental control parameters with headings and units. Each variable could be plotted as a function of time and data could be input directly from the keyboard, rather than from within the program itself. Simulation languages provide some of these facilities.

In designing simulation programs, however, the user must always consider the purpose of the study, the questions being asked and the predictions required to answer these questions. In the example of competition, the investigator might merely want to determine whether both species can co-exist in the chemostat. Results in *Tables 2–4* indicate that species 2 is being eliminated and no further information would be obtained by plotting these data. The examples above used one set of growth constants only, but if these were being varied then it would be useful to print them out. Similarly, if only one constant or initial condition was being varied, it would be easier to investigate its effect by entering the value from the keyboard rather than changing the program each time.

Computer programming is time consuming, even for experts, and the benefits of increased sophistication must be weighted against increased time, effort and cost. This is particularly true of graphics programming. Although it may provide much enjoyment and self-satisfaction and can impress onlookers, presentation of results in graphical form, in several colours, with complex axis labelling is of no use unless it aids in answering the initial questions set by a particular problem. Self indulgent programming must therefore always be avoided.

7. PARTIAL DIFFERENTIAL EQUATIONS

In many situations, parameters will vary with respect to two independent variables, requiring use of partial differential equations. Most commonly, the independent variables are time and distance. Two examples will be considered. The first is that of chemical diffusion which is an important feature of many biological systems. For example, growth of a bacterial colony on agar will depend on diffusion of metabolite through the agar and through the colony itself. In aquatic ecosystems, growth of certain organisms frequently occurs in bands at depths at which two nutrients diffusing from opposite directions are present at optimal combined concentrations. The reaction–diffusion theory for the establishment of spatial and positional information during development by biological structures is also based on diffusion of an activator and an inhibitor (20).

The process of diffusion was first quantified by Fick who stated that the rate of transfer (F) of a diffusing compound through unit area of a section is proportional to the concentration gradient across that area. This may be expressed mathematically as:

$$F = -D\partial^2 C/\partial x^2 \tag{16}$$

where C is the concentration of the diffusing compound and D is the diffusion coefficient. If D is assumed constant, the rate of change of C with time, t, in one dimension (along the x-axis) is given by Equation 17:

$$\partial C/\partial t = D\partial^2 C/\partial x^2 \tag{17}$$

The value of C at any distance, x, depends on both distance and time and changes in concentration must be described by partial differential equations. A treatment of the mathematics of diffusion, including numerical methods for solution of diffusion equations, is given in Crank (21).

A second example is the microbial conversion of substrate by a microbial population, biomass concentration m, in a column of particulate material. Nutrient medium is passed through the column at a constant rate and equations may be derived describing changes in substrate and biomass concentrations. The concentration of substrate, s, in an element of medium added to the top of the column will decrease as it passes down the column, due to microbial conversion. Concentration is therefore a function of distance, x. Conversion of s simultaneously leads to microbial growth, so m at each point in the column will also increase. The concentration of s in an element of medium added a short time later will also decrease with x, but at any particular distance the biomass concentration will differ from that of the previous time interval. The rate of conversion of s, and therefore its concentration, will be a function of both time and distance and cannot be described by ordinary differential equations.

This situation has been considered by Saunders and Bazin (22) in modelling nitrification in a glass bead column. Inorganic medium containing ammonium (s_1) is supplied to the top of the column and is converted by *Nitrosomonas* (biomass concentation m_1) to nitrite (s_2) which is further oxidized to nitrate (s_3) by *Nitrobacter* (biomass concentration m_2). This system may be modelled using partial differential equations to describe rates of change in s_1, s_2, s_3, m_1 and m_2:

$$\frac{\partial s_1}{\partial t} + \frac{f\,\partial s_1}{\partial x} = -k_1 m_1 s_1 \tag{18}$$

$$\frac{\partial s_2}{\partial t} + \frac{f\,\partial s_2}{\partial x} = -k_1 m_1 s_1 - k_2 m_2 s_2 \quad (19)$$

$$\frac{\partial s_3}{\partial t} + \frac{f\,\partial s_3}{\partial x} = k_2 m_2 s_2 \quad (20)$$

$$\frac{\partial m_1}{\partial t} = \frac{\mu_{m1} s_1 m_1}{Ks_1 + s_1} \quad (21)$$

$$\frac{\partial m_2}{\partial t} = \frac{\mu_{m2} s_2 m_2}{Ks_2 + s_2} \quad (22)$$

where k_1 and k_2 are rate constants for ammonium and nitrite conversion and f is flow-rate. Analytical solutions of these equations is only possible if certain simplifying assumptions are made and simulation of the complete model requires numerical approximation techniques.

Theoretical aspects of numerical approximation techniques for partial differential equations will now be described and these will be applied to a simple example of diffusion. An alternative technique for solution of the above system (Equations 18–22) will also be described.

7.1 **Theoretical aspects**

For reasons discussed in Section 6.1 the first aim in simulating a mathematical model should always be to obtain an analytical solution of that model. This is, however, much more difficult if the model consists of partial differential equations and approximation techniques are usually necessary. These are more difficult to implement than those for ordinary differential equations, in terms of the complexity of the computer programs, storage requirements, assessment of error generation and stability. Partial differential equations are classified as parabolic, hyperbolic and eliptic and the following discussion is for parabolic equations only. Consideration of the type of equation is important before carrying out numerical analysis and general texts on the subject (23–25) should be consulted.

7.1.1 *Explicit method*

Consider the diffusion in one dimension in a plane sheet, one end of which is maintained at a certain concentration of substance, C_0. This is analogous to, for example, the diffusion into a sediment of a component of the overlying water. Initially, the concentration of diffusant throughout the sheet is 0. Solution of the model requires calculations of C as a function of distance x at a series of time intervals, δt. Implementation of the approximation technique is facilitated by use of dimensionless variables, defined by the following equation;

$$\mathrm{X} = x/\mathrm{l} \qquad \mathrm{T} = Dt/\mathrm{l}^2 \qquad c = C/C_0 \quad (23)$$

where l is the total length being considered. X represents distance as a proportion of

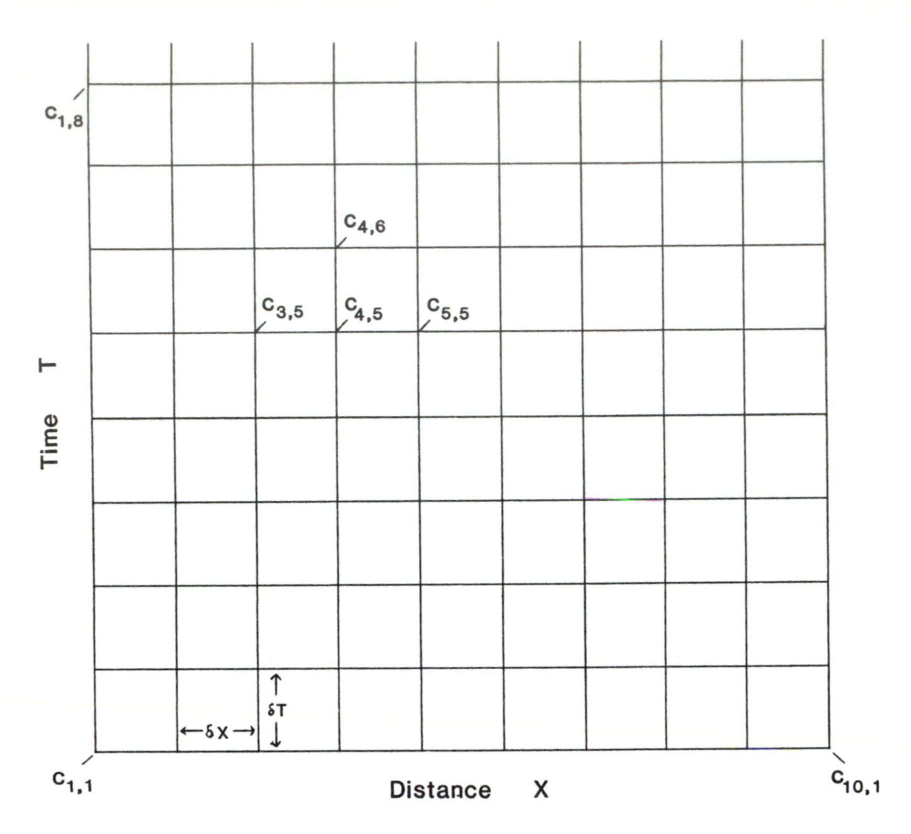

Figure 3. Two-dimensional rectangular grid representing values of concentration c as a function of distance and time. The distance (X) axis is divided into intervals of length X and the time axis into intervals of T. The value of c at distance i * X and time j * T is represented by $c_{i,j}$. The explicit method (see text) calculates values of c at time j + 1 from a knowledge of values at time j. $c_{4,6}$ would therefore be calculated by substitution of values $c_{3,5}$, $c_{4,5}$ and $c_{5,5}$ in Equation 30.

the total distance, c represents concentration of the substance as a proportion of its concentration at the boundary and time is represented by the dimensionless variable T, incorporating the diffusion coefficient D. Substitution of these values into Equation 17 yields the equation:

$$\partial c/\partial \mathrm{T} = \partial c^2/\partial \mathrm{T}^2 \tag{24}$$

The system can then be treated by dividing the plane sheet into layers of thickness δX and representing the solution as a two-dimensional rectangular grid with each grid containing the value of c at that particular distance and time interval. This is illustrated in *Figure 3* as a 10 × 8 grid with the concentration at a distance i * δX and time j * δT represented by $c_{i,j}$. Initially $c_{1,1} = 1$ and $c_{2,1}.......c_{10,1} = 0$. One end of the sheet i = 1 is maintained at constant concentration $c = 1$, so $c_{1,1}.....c_{1,10} = 1$. We therefore have the following boundary conditions:

$$c = 1, \quad 0>\mathrm{X}>1, \quad t\geq 0$$
$$c = 0, \quad \mathrm{X}>0, \quad t = 0$$

The solution to Equation 24 is obtained by applying a Taylor series expansion in the T direction, while keeping X constant, to obtain an expression for $\partial c/\partial t$, and then applying the same approximation in the X direction with T constant, given an expression for $\partial^2 c/\partial X^2$. The former is given by:

$$c_{i,j+1} = c_{i,j} + \partial T \frac{(\partial c)}{(\partial T)}_{i,j} + \tfrac{1}{2} (\partial T)^2 \frac{(\partial^2 c)}{(\partial T^2)}_{i,j} + \ldots \tag{25}$$

If all terms of order $(\delta T)^2$ and above are ignored, this simplifies to

$$\left(\frac{\partial c}{\partial T}\right)_{i,j} = \frac{c_{i,j+1} - c_{i,j}}{\delta T} \tag{26}$$

(remember that T contains the diffusion co-efficient D). Applying a Taylor series in the X direction with T constant gives:

$$c_{i+1,j} = c_{i,j} + \delta X \left(\frac{\partial c}{\partial X}\right)_{i,j} + \tfrac{1}{2} (\delta X)^2 \left(\frac{\partial^2 c}{\partial X^2}\right)_{i,j} + \ldots\ldots$$

$$c_{i-1,j} = c_{i,j} - \delta X \left(\frac{\partial c}{\partial X}\right)_{i,j} + \tfrac{1}{2} (\delta X)^2 \left(\frac{\partial^2 c}{\partial X^2}\right)_{i,j} + \ldots\ldots$$

summing these equations gives then

$$\left(\frac{\partial^2 c}{\partial X^2}\right)_{ij} = \frac{c_{i+1,j} - 2c_{i,j} + c_{i-1,j}}{(\delta X)^2} \tag{27}$$

An expression for $c_{i,j+1}$ is then obtained by substitution into Equation 17 using Equations 26 and 27 yielding:

$$c_{i,j+1} = c_{i,j} + r(c_{i-1,i,j} - 2c_{i,j} + c_{i+1,j}) \tag{28}$$

where $r = \delta T/(\delta X)^2$. If a value of $r = 0.5$ is chosen, Equation 28 reduces to:

$$c_{i,j+1} = (c_{i-1,j} + c_{i+1,j})/2$$

The arrangement of the four relevant values on the rectangular grid may be seen in *Figure 3*. Values at each distance increment (i) can be calculated for time increment j+1 from a knowledge of values at the previous time interval of j. This is therefore equivalent to the one-step technique described in Section 6.1.1 and is an example of an explicit method. A severe limitation on this technique is that for values of r greater than 0.5, the solution is unstable, in that errors increase without limit as t increases. This means that δT must be small compared with δX and a large number of small time steps are required by the method.

Iteration of solution of Equation 28 presents no problems if boundary conditions exist, that is if $c_{i,j}$ and $c_{10,j}$ of our example have known and independent values. Reflection at the boundary, as would occur at the edge of a Petri dish is also treated easily.

7.1.2 *Implicit methods*

As with the solution of ordinary differential equations, use of a greater number of values to estimate $c_{i,j+1}$ may be achieved by estimating $\partial^2 c/\partial x^2$ from information on the jth and (j+1)th time intervals. *Equation 20* is then replaced by:

$$-rc_{i-i,j+1} + (2+2r)c_{i,j+1} - rc_{i+1,j+1} = rc_{i-1,j} + (2-2r)c_{i,j} + rc_{i+1,j}$$

Again terms of the Taylor series of order greater than $(\delta X)^2$ have been neglected but the technique is stable for all values of r. Larger δT intervals may therefore be used, depending on the accuracy required, but much more computation is required at each time step. Calculation of c_{j+1} requires solution of a set of N simultaneous equations (where N is the number of distance increments). This technique is described in detail by Smith (25) and will not be considered here, but would generally be carried out using library routines designed for the task.

7.1.3 *Diffusion in two and three dimensions*

Both explicit and implicit methods may be applied to diffusion in two or three dimensions. For example, Equation 17 in two dimensions becomes

$$\frac{\partial c}{\partial T} = D\left(\frac{\partial^2 c}{\partial x^2}\right) + \left(\frac{\partial^2 c}{\partial y^2}\right)$$

where y represents the distance along the y-axis. A rectangular grid is now set up for each time step, containing values of $c_{i,j}$, where i represents the x-axis increment, j is now the y-axis increment and n is the time step. Equation 28 now becomes

$$\frac{c_{i,j,n+1} - C_{i,j,n}}{\partial t} = \frac{D}{(\delta x)^2}(C_{i-1,j,n}, - 2c_{i,j,n} + C_{i+1,j,n}) + \frac{D}{(\delta y)^2}(c_{i,j-1,n} - 2c_{i,j,n} + C_{i,j+1,n})$$

Stability now requires that $D\ (1/\delta x^2 + 1/\delta y^2)\delta t > 0.5$ so that much smaller values of δt are required and computation takes much longer. The problem is even greater for three dimensions. There may also be boundary problems at the corners of two- and three-dimensional grid systems.

7.2 **Practical aspects**

A comparison will be given here for simulation of a simple diffusion problem using three methods: (i) analytical solution, (ii) explicit method and (iii) FORTRAN program calling a NAG Library routine. The system to be considered is that of a plane sheet which has the following boundary conditions [from Smith (25)].

$c = 2X,$	$0 \leq X < 0.5,$	$T = 0$
$c = 2(1-X),$	$0.5 \leq X \leq 1,$	$T = 0$
$c = 0,$	$X = 0$ and $1,$	$T = 1$

The first two conditions define the gradient of c at $T = 0$, which increases linearly from $X = 0$ to $X = 0.5$ and then decreases linearly, with distance, to reach 0 at $X = 1$. Both edges of the sheet, $X = 0$ and $X = 1$, are kept at concentration $c = 0$. The initial gradient in c is therefore given by the first row of values in *Table 5*, which was obtained by analytical solution using Equation 29:

$$c = \frac{8}{\pi^2} \sum_{n=1}^{\infty} \frac{1}{n^2} \sin \tfrac{1}{2} n\pi \sin n \pi X \exp(-n^2\pi^2 T) \qquad (29)$$

Table 5. Variation in concentration, c, with distance (X) and time (T) during one-dimensional diffusion in a plane sheet with boundary conditions described in the text.

T	*Distance*										
	0.	*0.10*	*0.20*	*0.30*	*0.40*	*0.50*	*0.60*	*0.70*	*0.80*	*0.90*	*1.00*
0.025	0.	0.196	0.372	0.512	0.602	0.633	0.602	0.512	0.372	0.196	0.
0.050	0.	0.153	0.291	0.400	0.471	0.495	0.471	0.400	0.291	0.153	0.
0.075	0.	0.119	0.227	0.313	0.368	0.387	0.368	0.313	0.227	0.199	0.
0.100	0.	0.093	0.178	0.244	0.287	0.302	0.287	0.244	0.178	0.093	0.
0.125	0.	0.073	0.139	0.191	0.224	0.236	0.224	0.191	0.139	0.073	0.
0.150	0.	0.057	0.108	0.149	0.175	0.184	0.175	0.149	0.108	0.057	0.
0.175	0.	0.045	0.085	0.117	0.137	0.144	0.137	0.117	0.085	0.045	0.
0.200	0.	0.035	0.066	0.091	0.107	0.113	0.107	0.091	0.066	0.035	0.
0.225	0.	0.027	0.052	0.071	0.084	0.088	0.084	0.071	0.052	0.027	0.
0.250	0.	0.021	0.040	0.056	0.065	0.069	0.065	0.056	0.040	0.021	0.
0.275	0.	0.017	0.32	0.043	0.051	0.054	0.051	0.043	0.032	0.017	0.
0.300	0.	0.013	0.025	0.034	0.040	0.042	0.040	0.034	0.025	0.013	0.
0.325	0.	0.010	0.019	0.027	0.031	0.033	0.031	0.027	0.019	0.010	0.
0.350	0.	0.008	0.015	0.021	0.024	0.026	0.024	0.021	0.015	0.008	0.
0.375	0.	0.006	0.012	0.016	0.019	0.020	0.019	0.016	0.012	0.006	0.
0.400	0.	0.005	0.009	0.013	0.015	0.016	0.015	0.013	0.009	0.005	0.

Values are calculated by analytical solution of Equation 17 (Equation 31).

Values of c are calculated at 11 intervals of 0.1, from X = 0 − 1, over a series of time steps δT with a diffusion coefficient, D = 1. As T increases, the gradient c becomes non-linear and decreases at each depth. The final steady-state solution, T = ∞, is c = 0 throughout the x-axis, that is:

$$c = 0 \qquad 0 \leq X \leq 1, \qquad T = \infty$$

7.2.1 *Explicit method*

Program 5 simulates the model using the simple explicit method represented by Equation 28. The x-axis lies between 0 and 1 and is divided into 11 intervals of length 0.1 (DX). The time step size (DT) depends on DX and = R/DX^2. Simulation proceeds for TMAX/DT steps. The initial gradient in c is established by a DO LOOP which sets the values of the two-dimensional array C (I,J) where I is the X increment and J the T increment. Equation 28 is then solved with the required boundary conditions and T values are stored in array T. Results are then printed. Storage of all elements of the two-dimensional grid requires large array sizes for C and T and large storage capacity may be required for some situations. This is not usually a problem with one-dimensional diffusion, but diffusion in two and three dimensions requires two- and three-dimensional arrays.

Results are given in *Table 6* for R = 0.5 and 0.1. Comparison with *Table 5* demonstrates reasonably good approximation with most values within 3% and 2% of respective values of R. The largest discrepancies are either side of X = 0.5 and result from the initial discontinuity in the distribution of c at this point.

This method gives reasonable accuracy for this simple example and examples of its use in more complex biological systems are given in (20).

Table 6. Variation in c with X and T during one-dimensional diffusion using an explicit method for numerical approximation (*Program 5*) for values of (**a**) R = 0.5 and (**b**) R = 0.1.

T	*Distance*										
	0.	*0.10*	*0.20*	*0.30*	*0.40*	*0.50*	*0.60*	*0.70*	*0.80*	*0.90*	*1.00*
a											
0.	0.	0.200	0.400	0.600	0.800	1.000	0.800	0.600	0.400	0.200	0.
0.025	0.	0.189	0.365	0.513	0.614	0.651	0.614	0.513	0.365	0.189	0.
0.050	0.	0.153	0.293	0.404	0.476	0.501	0.476	0.404	0.293	0.153	0.
0.075	0.	0.121	0.230	0.316	0.372	0.391	0.372	0.316	0.230	0.121	0.
0.100	0.	0.094	0.180	0.247	0.291	0.306	0.291	0.247	0.180	0.094	0.
0.125	0.	0.074	0.140	0.193	0.227	0.239	0.227	0.193	0.140	0.074	0.
0.150	0.	0.058	0.110	0.151	0.178	0.187	0.178	0.151	0.110	0.058	0.
0.175	0.	0.045	0.086	0.118	0.139	0.146	0.139	0.118	0.086	0.045	0.
0.200	0.	0.035	0.067	0.092	0.109	0.114	0.109	0.092	0.067	0.035	0.
0.225	0.	0.028	0.053	0.072	0.085	0.089	0.085	0.072	0.053	0.028	0.
0.250	0.	0.032	0.041	0.057	0.066	0.070	0.066	0.057	0.041	0.022	0.
0.275	0.	0.017	0.032	0.044	0.052	0.055	0.052	0.044	0.032	0.017	0.
0.300	0.	0.013	0.025	0.035	0.041	0.043	0.041	0.035	0.025	0.013	0.
0.325	0.	0.010	0.020	0.027	0.032	0.033	0.032	0.027	0.020	0.010	0.
0.350	0.	0.008	0.015	0.021	0.025	0.026	0.025	0.021	0.015	0.008	0.
0.375	0.	0.006	0.012	0.017	0.019	0.020	0.019	0.017	0.012	0.006	0.
0.400	0.	0.005	0.009	0.013	0.015	0.016	0.015	0.013	0.009	0.0	
b											
0.	0.	0.200	0.400	0.600	0.800	1.000	0.800	0.600	0.400	0.200	0.
0.025	0.	0.188	0.375	0.500	0.625	0.625	0.625	0.500	0.375	0.188	0.
0.050	0.	0.156	0.283	0.410	0.459	0.508	0.459	0.410	0.283	0.156	0.
0.075	0.	0.116	0.232	0.304	0.375	0.375	0.375	0.304	0.232	0.116	0.
0.100	0.	0.095	0.172	0.248	0.278	0.307	0.278	0.248	0.172	0.095	0.
0.125	0.	0.070	0.140	0.184	0.227	0.227	0.227	0.184	0.140	0.070	0.
0.150	0.	0.057	0.104	0.150	0.168	0.186	0.168	0.150	0.104	0.057	0.
0.175	0.	0.043	0.085	0.111	0.138	0.138	0.138	0.111	0.085	0.043	0.
0.200	0.	0.035	0.063	0.091	0.102	0.113	0.102	0.091	0.063	0.035	0.
0.225	0.	0.026	0.051	0.067	0.083	0.083	0.083	0.067	0.051	0.026	0.
0.250	0.	0.021	0.038	0.055	0.062	0.068	0.062	0.055	0.038	0.021	0.
0.275	0.	0.016	0.031	0.041	0.050	0.050	0.050	0.041	0.031	0.016	0.
0.300	0.	0.013	0.023	0.033	0.037	0.041	0.037	0.033	0.023	0.013	0.
0.325	0.	0.009	0.019	0.025	0.031	0.031	0.031	0.025	0.019	0.009	0.
0.350	0.	0.008	0.014	0.020	0.023	0.025	0.023	0.020	0.014	0.008	0.
0.375	0.	0.006	0.011	0.015	0.018	0.018	0.018	0.015	0.011	0.006	0
0.400	0.	0.005	0.008	0.012	0.014	0.015	0.014	0.012	0.008	0.005	0.

7.2.2 *Use of the NAG Library*

Program 6 demonstrates use of a NAG library routine which is one of several designed to solve parabolic and elliptic equations with specified boundary conditions. Routine DO3PAF solves parabolic equations of the type

$$\frac{\partial \mu}{\partial T} = X^{-M} \frac{\partial}{\partial X} \left[X^{M} G(X,T,U) \frac{\partial U}{\partial X} \right] + F \left[X,T,U, \frac{\partial U}{\partial X} \right]$$

Table 7. Variation in c with X and T using a NAG Library routine (*Program 6*).

The relative accuracy in the time integration is 0.1D−04

T	*Distance*										
	0.	*0.10*	*0.20*	*0.30*	*0.40*	*0.50*	*0.60*	*0.70*	*0.80*	*0.90*	*1.00*
0.	0.	0.200	0.400	0.600	0.800	1.000	0.800	0.600	0.400	0.200	0.
0.025	0.	0.188	0.364	0.514	0.616	0.652	0.616	0.514	0.364	0.188	0.
0.050	0.	0.154	0.293	0.405	0.477	0.503	0.477	0.405	0.293	0.154	0.
0.075	0.	0.121	0.230	0.317	0.373	0.392	0.373	0.317	0.230	0.121	0.
0.100	0.	0.095	0.180	0.248	0.292	0.307	0.292	0.248	0.180	0.095	0.
0.125	0.	0.074	0.141	0.194	0.229	0.240	0.229	0.194	0.141	0.074	0.
0.150	0.	0.058	0.111	0.152	0.179	0.188	0.179	0.152	0.111	0.058	0.
0.175	0.	0.046	0.087	0.119	0.140	0.147	0.140	0.119	0.087	0.046	0.
0.200	0.	0.036	0.068	0.093	0.110	0.115	0.110	0.093	0.068	0.036	0.
0.225	0.	0.028	0.053	0.073	0.085	0.090	0.086	0.073	0.053	0.028	0.
0.250	0.	0.022	0.042	0.057	0.067	0.071	0.067	0.057	0.042	0.022	0.
0.275	0.	0.017	0.033	0.045	0.053	0.055	0.053	0.045	0.033	0.017	0.
0.300	0.	0.013	0.025	0.035	0.041	0.043	0.041	0.035	0.025	0.013	0
0.325	0.	0.010	0.020	0.027	0.032	0.034	0.032	0.027	0.020	0.010	0.
0.350	0.	0.008	0.016	0.022	0.025	0.027	0.025	0.022	0.016	0.008	0.
0.375	0.	0.006	0.012	0.017	0.020	0.021	0.020	0.017	0.012	0.006	0.
0.400	0.	0.005	0.010	0.013	0.015	0.016	0.015	0.013	0.010	0.005	0.

Equation 17 is an example of such an equation with U = c, M = 0, F = 0 and G = D. Boundary conditions are represented by

$$\mathrm{P(T)}c + \mathrm{Q(T)}\frac{\partial c}{\partial \mathrm{X}} = \mathrm{R(T,}c\mathrm{)}$$

and values of P, Q and R must be provided for X at points A and B. For our example, the boundary conditions are not a function of X and $\partial c/\partial$X, and therefore Q(T) = 0. At both X = A and X = B, c = 0 and is not a function of time. The coefficients at X = A and B are therefore P(T) = 1 and R(T,c) = 0.

The program structure is similar to that of *Program 3*, with a subroutine, DO3PAF, called from the main program with several arguments. The coordinate system is defined by M and in this example is cartesian (M = 0). A and B are the boundary values for X (A = 0, B = 1) and T and TOUT are initial and final values for T. The one-dimensional array C contains concentrations at values of distance along the x-axis which is divided into NPTS (11) sections. The accuracy is controlled by estimating local error and adjusting the integration step size accordingly as defined by ACC. Other arguments concern work space required by the subroutine and failure diagnostics.

After provision of values for subroutine arguments the initial gradient is established in a DO LOOP. Results are printed after each subroutine call.

Two user-defined subroutines must be provided. PDEF supplies values for F (= 0) and G (= D = 1). Subroutine BNDY sets boundary conditions by providing values for P, Q, and R (see above) at X = 0 (IBND = 0) and X = 1 (IBND = 1).

Results are given in *Table 7* and, with the value of ACC = 0.001, approximations are generally within 2% of the analytical solution. Both methods therefore give reasonable accuracy for this simple example.

7.2.3 *Simulation of continuous flow particulate reactor*

A second example of the use of partial differential equations is the sequential microbial conversion of ammonium to nitrate in a column of particulate material, as described above by Equations 18–22. Analytical solution of this model is possible if simplifying assumptions are made, but this reduces the range of substrate concentrations over which it may be applied. Solutions of Equations 18–22 by library routines have not yet been attempted but an alternative approach (26) is to divide the column into a series of compartments of equal width. Each compartment is considered homogeneous with respect to biomass and substrate concentrations and can therefore be described by ordinary differential equations. The length of each compartment is chosen such that the time taken for the medium front to pass through the compartment is constant. It will therefore depend on the total length of the column and flow-rate, faster rates giving fewer and larger compartments. At each time interval, changes in biomass and substrate concentrations in each compartment may be determined by solution of ordinary differential equations as described in Section 6, using any suitable growth model. At the end of the time interval non-fixed components are transferred to each compartment to the compartment immediately below. Fixed components (for example attached biomass) are not transferred. The column may therefore be considered as a series of linked chemostats, but with no transfer of biomass.

Details of the simulation results may be found in the original publication (26). The technique allows investigation of transient conditions such as change in flow-rate and can incorporate any suitable model for growth and substrate conversion formulated in terms of ordinary differential equations. It can also simulate discontinuities within the system if required.

7.3 **Comparison of techniques**

Library software routines for solution of partial differential equations have similar advantages to those for ordinary differential equations, in that they are extensively tested and provide greater facilities for error analysis, controlled accuracy and failure diagnostics. Not all equations will be capable of solution and care must therefore be taken in their use. Depending on the type of equation, use of techniques is restricted to one or, more rarely, two and three space dimensions and in the routine used above boundary conditions must be supplied for all X values at T = 0 and at X = 0 and X = T (A and B) for all subsequent times. Problems also exist for certain boundary corners and when singularities and discontinuities occur. Finite difference techniques written by the user, while not as rigorous, may often be preferable to the microbiologist with limited mathematical training. They provide better visualization of changes in the behaviour of a system and permit a degree of complexity which is difficult and often impossible to obtain in the form of partial differential equations which describe continuous changes. They often, therefore, constitute computer models, rather than purely mathematical models. The choice of technique depends even more than for ordinary differential equations on the particular system being studied, the types of equations within the model and the programming time and expertise available.

8. CONCLUSIONS

Mathematical modelling is a powerful tool in the description and analysis of microbial systems and processes. Most models are formulated as a set of ordinary or partial differential equations which must be solved to provide predictions of system behaviour. Analytical solutions must always be sought but this is not usually possible because of the complexity of biological systems, and consequently of the mathematical models. Simulation therefore requires numerical approximation techniques. These are well established for ordinary differential equations, and simulation languages and library routines provide excellent facilities for their solution. The former are ideal for less experienced users with uncomplicated models but some models may require specific facilities provided only by library routines. Experienced programmers may prefer the greater flexibility and control associated with the latter. Simulations are usually carried out on mainframe computers but increasing speed and storage capacity will soon allow efficient simulation on microcomputers. Solution of partial differential equations presents greater problems but is possible by user-defined, problems-specific approximation techniques and by library routines. They are necessarily more complex and advice should be sought at an early stage if there is doubt regarding the method to be used. It is also more difficult to quantify errors and stability.

In all stimulation studies it is necessary to carry out extensive tests. Programming errors must obviously be eliminated first and step size should then be varied to determine the significance of local and global errors. Overflow conditions must also be tested for and eliminated.

The development of approximation techniques, and of computer algorithms for high-speed digital computers permits the simulation of models which was inconceivable 25 years ago. The techniques described in this chapter permit realization of the enormous potential of mathematical modelling and allow its application by users with little previous experience of computing.

9. REFERENCES

1. Jason,A.C. (1983) *Antonie von Leeuwenhoek,* **19**, 513.
2. Roels,J.A. and Kossen,N.W.F. (1978) *Progr. Ind. Microbiol.*, **14**, 95.
3. Harvey,J.D. (1983) In *Mathematics in Microbiology*. Bazin,M.J. (ed.), Academic Press, London, p.1.
4. Segel,L.A. (ed.) (1980) *Mathematical Models in Molecular and Cell Biology*. Cambridge University Press, Cambridge.
5. Bazin,M.J. (ed.) (1982) *Microbial Population Dynamics*. CRC Press, Inc., Boca Raton, Florida.
6. Bazin,M.J. and Prosser,J.I. (eds) (1987) *Physiological Models in Microbiology*. CRC Press Inc., Boca Raton, Florida.
7. Bull,A.T. and Slater,J.H. (eds) (1982) *Microbial Interactions and Communities*. Academic Press, London.
8. Kleinsreuer,C. and Poweigha,T. (1984) *Adv. Biochem. Eng.*, **30**, 91.
9. Votruba,J. (1986) In *Overproduction of Microbial Metabolites*. Variek,Z. and Hostalck,Z. (eds), Butterworths, London.
10. Harder,A. and Roels,J.A. (1982) *Adv. Biochem. Eng.*, **21**, 55.
11. Saunders,P.T. (1980) *An Introduction to Catastrophe Theory*. Cambridge University Press, Cambridge.
12. Hildebrand,F.B. (1974) *Introduction to Numerical Analysis*. McGraw-Hill Book Company, New York.
13. Hall,G. and Watt,J.M. (eds) (1976) *Modern Numerical Methods for Ordinary Differential Equations*. Oxford University Press, Oxford.
14. Korn,G.A. and Wait,J.V. (1978) *Digital Continuous-System Simulation*. Prentice-Hall, Inc., New Jersey.
15. *NAG FORTRAN Library Manual, Mark II* (1984) Numerical Algorithms Group, Oxford.

16. *6000 Series MIMIC Digital Simulation Language Reference Manaul* (1972) Control Data Corporation Systems, Sunnyvale, California.
17. Maryanski,F. (1980) *Digital Computer Simulation*. Hayden Book Company, Inc., New Jersey.
18. Heijnen,J.J., Roels,J.A. and Stouthamer,A.H. (1979) *Biotechnol. Bioeng.*, **21**, 2175.
19. Clore,G.M. (1983) In *Computing in Biological Science*. Geisow,M.J. and Barrett,A.N. (eds), Elsevier Biomedical Press, Amsterdam, p. 313.
20. Meinhardt,H. (1982) *Models of Biological Pattern Formation,* Academic Press, London and New York.
21. Crank,J. (1975) *The Mathematics of Diffusion*. Clarendon Press, Oxford.
22. Saunders,P.T. and Bazin,M.J. (1973) *Soil Biol. Biochem.*, **5**, 545.
23. Mitchell,A.R. (1969) *Computational Methods in Partial Differential Equations*. John Wiley & Sons, London.
24. Mitchell,A.R. and Wait,R. (1967) *The Finite Element Method in Partial Differential Equations*. John Wiley & Sons, London.
25. Smith,G.D. (1978) *Numerical Solution of Partial Differential Equations: Finite Difference Methods*. Oxford Unviersity Press, Oxford.
26. Prosser,J.I. and Gray,T.R.G. (1977) *J. Gen. Microbiol.*, **102**, 119.

10. APPENDIX

Program 1. Simulation of the logistic equation (Equation 4) with μ_m = 0.5/h, k = 0.01 and x_0 = 1.0, using the analytical solution (Equation 12), Euler's method and a Runge–Kutta technique. The program is written in BASIC for a BBC model B microcomputer.

Note: BBC BASIC permits the use of procedures. To convert the program to other versions of BASIC, procedures can be replaced by subroutines, for example:

	BBC BASIC		Other BASICS
60	PROC INIT	60	GOSUB 150
150	DEF PROC INIT	150	REM Initialization Routine
230	END PROC	230	RETURN

```
 10 MODE3
 20 REM   Dimension arrays for Runge-Kutta method and for output of results
 30 DIM Y(1),F(1),S(1),P(1),T(50),X1(50),X2(50),X3(50)
 40 REM   Call procedures for intialisation, analytical and numerical solutions
 50 REM          and output of results
 60 PROCINIT
 70 PROCANALYT
 80 PROCEULER
 90 PROCRUNGE
100 PROCOUTPUT
110 STOP
120 END
130
150 DEF PROCINIT
160 REM   Provide values for initial conditions, growth constants and TMAX
170 X0=1. :X1(1)=X0 :X2(1)=X0 :X3(1)=X0 :Y(1)=X0
180 UM=.5 :K=.01
190 H=0.5 :T=0. :TMAX=25. :TINT=1. :TN=INT(TMAX/TINT)+1
200 FOR I = 1 TO TN
210   T(I)=(I-1)*TINT
220 NEXT I
230 ENDPROC
240
260 DEF PROCANALYT
270 REM   Analytical solution
280 FOR I = 1 TO TN
290   X1(I) = X0/((1-K*X0)*EXP(-UM*T(I))+K*X0)
300 NEXT I
310 ENDPROC
320
340 DEF PROCEULER
350 REM   Numerical solution using Euler method
```

PROGRAM 1 (continued):

```
360 I1=2 :X=X0
370 FOR I = 1 TO TMAX/H
380   X = X + H * UM*X*(1-K*X) :T=T+H
390   IF(ABS(T-T(I1))>H/2) THEN 410
400   X2(I1)=X  :I1=I1+1
410 NEXT I
420 ENDPROC
430
450 DEF PROCRUNGE
460 REM   Numerical solution using Runge-Kutta method
470 I1=2 :N=1 :F1=1 :T=0.
480 M=0
490 FOR I = 1 TO TMAX/H
500   PROCRK
510   IF F1<>1 THEN 540
520   F(1) = UM*Y(1)*(1-K*Y(1))
530   GOTO 500
540   IF(ABS(T-T(I1))>H/2) THEN 560
550   X3(I1)=Y(1) :I1=I1+1
560 NEXT I
570 ENDPROC
580
600 DEF PROCRK
610 REM   Calculate K0, K1, K2, K3 for Runge-Kutta method
620 M=M+1
630 ON M GOTO 640,660,710,760,810
640 F1=1
650 ENDPROC
660 FOR J=1 TO N
670   S(J)=Y(J) :P(J)=F(J) :Y(J)=S(J)+0.5*H*F(J)
680 NEXT J
690 T=T+0.5*H :F1=1
700 ENDPROC
710 FOR J=1 TO N
720   P(J)=P(J)+2*F(J) :Y(J)=S(J)+0.5*H*F(J)
730 NEXT J
740 F1=1
750 ENDPROC
760 FOR J=1 TO N
770   P(J)=P(J)+2*F(J) :Y(J)=S(J)+H*F(J)
780 NEXT J
790 T=T+0.5*H :F1=1
800 ENDPROC
810 FOR J=1 TO N
820   Y(J)=S(J)+(P(J)+F(J))*H/6
830 NEXT J
840 M=0 :F1=0
850 ENDPROC
860
880 DEF PROCOUTPUT
890 REM   Output results from analytical and numerical solutions
900 PRINT "STEP SIZE = ";H :PRINT
910 PRINT "TIME";TAB(14);"BIOMASS CONCENTRATION";TAB(55);"%ERROR"
920 PRINT TAB(7);"ANALYTICAL"; TAB(23);"EULER";TAB(32);"RUNGE-KUTTA";
930 PRINT TAB(48);"EULER";TAB(62);"RUNGE-KUTTA"
940 FOR I=1 TO TN
950   X2=(X1(I)-X2(I))*100/X1(I) :X3=(X1(I)-X3(I))*100/X1(I)
960   PRINT TAB(2);T(I);TAB(7);X1(I);TAB(20);X2(I);TAB(32);X3(I);
970   PRINT TAB(46);X2;TAB(62);X3
980 NEXT I
990 ENDPROC
```

Program 2. Simulation of competition by two microbial populations by single limiting nutrient in a chemostat (Equation 13–15). The program is written in BASIC for a BBC model B microcomputer and uses a Runge–Kutta technique for solution of differential equations.

```
 10 MODE3
 20 REM   Dimension arrays for Runge-Kutta method and for output of results
 30 DIM Y(3),F(3),S(3),P(3),T(50),X1(50),X2(50),X3(50)
 40 PROCINIT
 50 PROCRUNGE
 60 PROCOUTPUT
 70 STOP
 80 END
 90 REM
100 DEF PROCINIT
110 REM   Provide values for growth constants, initial conditions and
120 REM         experimental control parameters
130 X1(1)=1. :X2(1)=1. :X3(1)=0. :Y(1)=1. :Y(2)=1. :Y(3)=0.
140 UM1=0.5 :UM2=0.25 :KS1=5.0 :KS2=1.0 :Y1=0.4 :Y2=0.6
150 D=0.2 :SR=50.
160 H=0.5 :T=0. :TMAX=100. :DT=5. :TN=INT(TMAX/DT)+1
170 FOR I = 1 TO TN
180   T(I)=(I-1)*DT
190 NEXT I
200 ENDPROC
210 REM
220 DEF PROCRUNGE
230 REM   Solve differential equations using Runge-Kutta method
240 I1=2 :N=3 :F1=1
250 M=0
260 FOR I = 1 TO TMAX/H
270   PROCRK
280   IF F1 <> 1 THEN 350
290   U1 = UM1*Y(3)/(KS1+Y(3))
300   U2 = UM2*Y(3)/(KS2+Y(3))
310   F(1) = U1*Y(1)-D*Y(1)
320   F(2) = U2*Y(2)-D*Y(2)
330   F(3) = D*SR-U1*Y(1)/Y1-U2*Y(2)/Y2-D*Y(3)
340   GOTO 270
350   IF(ABS(T-T(I1))>H/2)THEN 370
360   X1(I1)=Y(1) :X2(I1)=Y(2) :X3(I1)=Y(3) :I1=I1+1
370 NEXTI
380 ENDPROC
390 REM
400 DEF PROCRK
410 Calculate K0, K1, K2 and K3 for Runge-Kutta method
420 M=M+1
430 ON M GOTO 440,450,500,550,600
440 F1=1 :ENDPROC
450 FOR J=1 TO N
460   S(J)=Y(J) :P(J)=F(J) :Y(J)=S(J)+0.5*H*F(J)
470 NEXT J
480 T=T+0.5*H :F1=1
490 ENDPROC
500 FOR J=1 TO N
510   P(J)=P(J)+2*F(J) :Y(J)=S(J)+0.5*H*F(J)
520 NEXT J
530 F1=1
540 ENDPROC
550 FOR J=1 TO N
560   P(J)=P(J)+2*F(J) :Y(J)=S(J)+H*F(J)
570 NEXT J
580 T=T+0.5*H :F1=1
590 ENDPROC
600 FOR J=1 TO N
610   Y(J)=S(J)+(P(J)+F(J))*H/6
```

PROGRAM 2 (continued):

```
620 NEXT J
630 M=0 :F1=0
640 ENDPROC
660 REM
670 DEF PROCOUTPUT
680 PRINT "STEP SIZE = ";H :PRINT
690 PRINT "TIME";TAB(15);"X1";TAB(30);"X2";TAB(45);"S"
700 FOR I=1 TO TN
710   PRINT TAB(1);T(I);TAB(10);X1(I);TAB(25);X2(I);TAB(40);X3(I)
720 NEXT I
730 ENDPROC
```

Program 3. Simulation of competition models (Equations 13–15). The program is written in FORTRAN and uses a Runge–Kutta–Merson technique evaluated by a NAG Library subroutine.

```
      DOUBLE PRECISION H,TMAX,TOL,T,DT
      INTEGER I,IFAIL,IR,J,N
      DOUBLE PRECISION W(3,7),Y(3)
      EXTERNAL FCN,OUTPUT
      COMMON TMAX,H,I,DT
C     Provide values for arguments of integration subroutine
      N=3
      IR=0
      TOL = .00001
      T=0.
      TMAX=100.
      Y(1)=1.
      Y(2)=1.
      Y(3)=0.
      DT=5.D0
      IFAIL=1
      WRITE(6,9999)
C     Call integration subroutine
      CALL D02BBF(T,TMAX,N,Y,TOL,IR,FCN,OUTPUT,W,IFAIL)
 9999 FORMAT(1H ,"TIME",10X,"X1",9X,"X2",10X,"S")
      STOP
      END
      SUBROUTINE FCN(T,Y,F)
      DOUBLE PRECISION T,UM1,UM2,KS1,KS2,Y1,Y2,D,SR,U1,U2
      DOUBLE PRECISION Y(3),F(3)
C     Provide values for growth constants and experimental conditions
      DATA UM1/0.5D0/,UM2/0.25D0/,KS1/5.0D0/,KS2/1.0D0/,Y1/0.4D0/,Y2/0.6D0/
      DATA D/0.2D0/,SR/50.D0/
C     Calculate functions for differential equations
      U1=UM1*Y(3)/(KS1+Y(3))
      U2=UM2*Y(3)/(KS2+Y(3))
      F(1)=U1*Y(1)-D*Y(1)
      F(2)=U2*Y(2)-D*Y(2)
      F(3)=D*SR-U1*Y(1)/Y1-U2*Y(2)/Y2-D*Y(3)
      RETURN
      END
      SUBROUTINE OUTPUT(T,Y)
C     Output of results
      DOUBLE PRECISION T,H,TMAX,DT
      DOUBLE PRECISION Y(3)
      INTEGER I
      COMMON TMAX,H,I,DT
      WRITE(6,9998)T,(Y(J),J=1,3)
      T=T+DT
      I=I-1
      RETURN
 9998 FORMAT(1H ,F5.1,2X,3F12.7)
      END
```

Program 4. Simulation of competition model (Equations 13–15). The program is written in MIMIC.

```
C                       Define constants for model simulation
                        CON(UM1,UM2,KS1,KS2,Y1,Y2)
                        CON(X10,X20,S0)
                        CON(SR,TMAX,DT)
C                       Define parameters for simulation
                        PAR(D)
C                       Define equations
              U1        UM1*S/(KS1+S)
              U2        UM2*S/(KS2+S)
              X1        INT(U1*X1-D*X1,X10)
              X2        INT(U2*X2-D*X2,X20)
              S         INT(D*SR-U1*X1/Y1-U2*X2/Y2-D*S,S0)
C                       Terminate simulation
                        FIN(T,TMAX)
C                       Output results
                        HDR(T,X1,X2,S)
                        OUT(T,X1,X2,S)
                        PLO(T,X1,X2)
                        PLO(S)
                        END
C                       Provide data for constants and parameters
 .5         .25         5.          1.          0.4         0.6
 1.         1.          0.
 50.        98.         5.
 .2
 .05
```

Program 5. Program for simulation of one-dimensional diffusion in a plane using an explicit method.

```
      REAL C(12,500),T(500)
C     Set up two-dimensional grid
      R=0.1
      T(1)=0.
      NX=11
      DX=1/(FLOAT(NX)-1)
      DT=R*DX^2
      NT=401
C     Set up intial gradient
      DO 110 I=2,NX-1
      X=(I-1)*DX
      IF(I.GT.NX/2)GOTO 120
      C(I,1)=2*X
      GOTO 110
  120 C(I,1)=2*(1-X)
  110 CONTINUE
C     Solve equation
      DO 130 J=1,NT
      T(J+1)=DT*J
      C(1,J)=0.
      C(NX,J)=0.
        DO 140 I=2,NX-1
        C(I,J+1)=C(I,J)+R*(C(I-1,J)-2*C(I,J)+C(I+1,J))
  140   CONTINUE
  130 CONTINUE
C     Output results
      WRITE(6,9999)
      WRITE(6,9998)(DX*(I-1),I=1,NX)
      WRITE(6,9997)(T(J),(C(I,J),I=1,NX),J=1,NT,25)
 9999 FORMAT(1H ,35X,"DISTANCE")
 9998 FORMAT(1H ,"  T  ",11F6.2//)
 9997 FORMAT(1H ,F5.3,1X,11F6.3)
      STOP
      END
```

Program 6. Program for simulation of one-dimensional diffusion in a plane using a NAG Library subroutine.

```
      DOUBLE PRECISION A,ACC,B,T,DT,TOUT
      INTEGER I,IFAIL,IN,IND,INP,IT,IWK,M,NPTS
      DOUBLE PRECISION C(20),WORK(400),X(11),DX,DT
      EXTERNAL BNDY,PDEF
C     Provide values for subroutine arguments
      IWK = 400
      M = 0
      A = 0.D+0
      B = 1.D+0
      ACC = 0.1D-4
      WRITE (6,9999) ACC
      T = 0.D+0
      IND = 0
      NPTS = 11
      DX=0.1D+0
      DT=0.025
C     Establish initial gradient
        DO 110 I=2,NPTS-1
        IF(I.GT.NPTS/2)GOTO 100
        C(I)=(I-1)*DX*2
        GOTO 110
  100   C(I)=2*(1-(I-1)*DX)
  110   CONTINUE
      C(1)=0.D+0
      C(NPTS)=0.D+0
      WRITE(6,9998)
      WRITE(6,9997)(DX*(I-1),I=1,NPTS)
        DO 120 IT=1,17
        TOUT=DT*(IT-1)
        IFAIL = 1
C     Call subroutine
        CALL D03PAF(M,A,B,T,TOUT,C,NPTS,ACC,WORK,IWK,IND,IFAIL)
C     Output results
        WRITE(6,9996)TOUT,(C(I),I=1,NPTS)
  120   CONTINUE
 9999 FORMAT(50H THE RELATIVE ACCURACY IN THE TIME INTEGRATION IS ,
     & D10.1//)
 9998 FORMAT(1H ,35X,"DISTANCE")
 9997 FORMAT(1H ,"  T  ",11F6.2//)
 9996 FORMAT(1H ,F5.3,1X,11F6.3)
      STOP
      END
      SUBROUTINE PDEF(X,T,CX,DCX,F,G)
C     Supply values for F and G
      DOUBLE PRECISION DCX,F,G,T,CX,X,D
      DATA D/1.D+0/
      F = 0.D+0
      G = D
      RETURN
      END
      SUBROUTINE BNDY(T,CX,IBND,P,Q,R)
C     Set boundary conditions
      DOUBLE PRECISION P, Q, R, T, CX
      INTEGER IBND
      IF (IBND.EQ.1) GO TO 100
      P = 1.D+0
      Q = 0.D+0
      R =0.D+0
      GO TO 110
  100 P = 1.D+0
      Q = 0.D+0
      R =0.D+0
  110 CONTINUE
      RETURN
      END
```

CHAPTER 7

Computers in clinical microbiology

KEITH A.V.CARTWRIGHT

1. INTRODUCTION

The use of microprocessors in clinical microbiology laboratories has a comparatively short history. Early attempts to handle data made use of punched cards (1) and paper tapes (2) but these were rapidly superseded. Data storage capacities and microprocessor power and flexibility have advanced by quantum jumps over the last 10 years and costs in relative terms have crumbled. Now there are few clinical microbiology laboratories which do not make some use of microprocessors to control automated equipment, in data capture, for word processing or in data recording and manipulation. The size and sophistication of many of these systems is impressive; enthusiasts cite the many advantages which 'computerization' confers on the laboratory and on the users; sceptics point to the high cost, potential for breakdown and failure to translate the potential into reality.

Why are computers needed? The possible benefits are considerable (*Figure 1*). Despite the obvious attractions, clinical microbiologists have been slower to undertake complete computerization of their data handling than colleagues in other pathology specialities, though their interest in the new technology is as great. Computerization of a microbiology laboratory presents greater difficulties than for any of the other pathology disciplines.

The problems of data entry are common to all; most laboratories receive their specimens accompanied by hand-written requests; the data include a registration number, date and nature of request, nature of specimen submitted, patient's name and date of birth, referring consultant, address for final report and sometimes brief clinical details. If the laboratory has a computer these pieces of information must be transferred to an electronic database. Various methods including the extensive use of codes (3), bar codes and optical mark readers (4) have been employed in an effort to minimize the work involved but the task remains formidable. The introduction onto hospital wards of computer terminals linked to the laboratory so that requests could be made directly by physicians would eliminate the need for copying of the data resulting in time-saving and removal of transcription errors. Ultimately the linking of family practitioners and other users to the laboratory, either directly or via the telephone network, would obviate this unnecessary work altogether.

Though the workload in clinical chemistry and haematology laboratories is much greater than in microbiology, results in these disciplines are generated principally in numeric form, often transmitted directly from an analyser to a linked computer. Few textual reports are issued to users and thus these departments are able to minimize the data handling generated by their large specimen loads. Histopathology departments,

Reduction of drudgery - typing, photocopying, sorting and filing of reports

Better service to users - legible reports

faster response to telephone enquiry

direct links to users possible

Better quality microbiology - access to previous reports

reduction of transcription errors

standardisation of test results

e.g. colorimetric

Containment of increasing workload within restricted staff budget.

Improved communications - other labs, hospitals, libraries etc.

Management data - staff, workload statistics, Korner data

Epidemiology - local and national

Word processing

Stock control - scope for collaborative purchasing policy

Patient billing

Figure 1. Advantages of computerization.

Costs - capital and recurrent

Slow data entry

Hidden consumption of human resources in maintenance and programming

May be slow (or perceived to be)

Inflexibility - in data entry, handling or output

Unreliability (also, dependence is induced)

Inherently out of date

Figure 2. Disadvantages of computerization.

by contrast, issue mostly long textual reports but handle modest numbers of specimens.

Clinical microbiology laboratories have the worst of both worlds. There is a big specimen load often about half that received by chemistry and haematology departments, but results contain substantial amounts of text including interpretive comment. Add

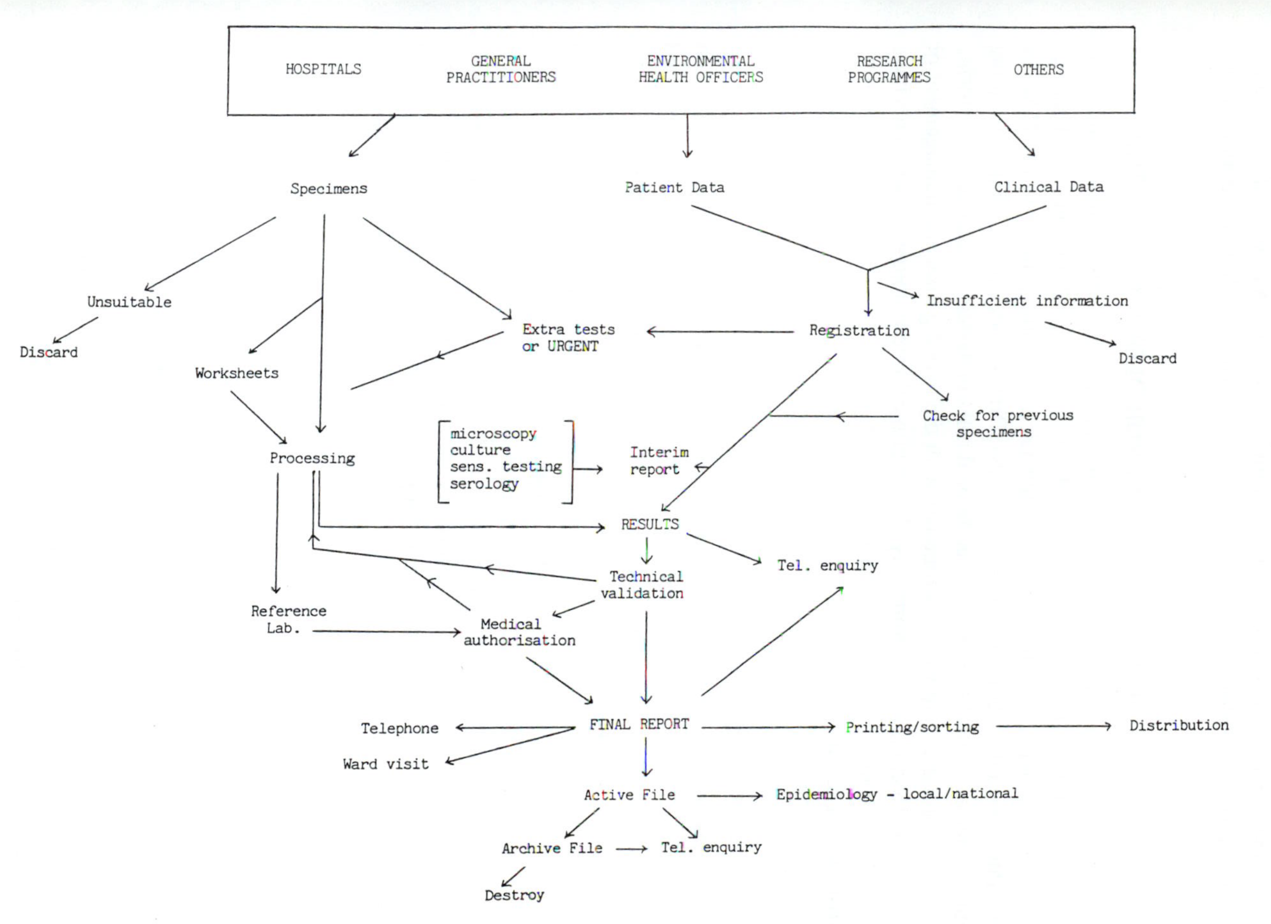

Figure 3. Simplified flowchart of activities within a clinical microbiology laboratory.

to this the slow development of automation within microbiology laboratories and it is easy to see why full computerization has progressed only slowly. A number of total microbiology laboratory systems have been developed, some as part of pan-pathology systems and some as autonomous microbiology systems. Whilst conferring some benefits, all have their disadvantages (*Figure 2*) and further evolution is both desirable and inevitable.

2. ORGANIZATION OF A CLINICAL MICROBIOLOGY LABORATORY

The complexity of clinical microbiology computing systems reflects the organization of these laboratories (*Figure 3*), which, in the UK may be administered and funded either by the National Health Service through District Health Authorities (DHAs) or by the Public Health Laboratory Service (PHLS). PHLS laboratories, whilst providing a routine microbiological service, have additional roles in the surveillance and monitoring of infectious diseases and vaccines together with the investigation and management of outbreaks of infection in the community. Some clinical microbiology departments have close links with university departments of microbiology. Departmental funding may be from any of the above or, more commonly, costs may be shared and must therefore be correctly apportioned.

2.1 **Throughput of specimens**

Most specimens arrive during the day and are processed in batches. A small proportion arrive and are dealt with outside normal working hours. Specimens may be submitted by hospital doctors, general practitioners (GPs), environmental health officers, occupational health departments, other laboratories or as part of research projects or quality control schemes. Requests may range from bacteriological or virological isolation through microscopical examination or serological testing to the assay of anti-microbial drug concentrations. The range of tests undertaken on particular specimens will depend on the requests made by the referring doctor and on the clinical details supplied; tests may be added to or amended by the laboratory staff and may take anything from minutes to a few weeks to complete, though most reports are issued within 48 h of the arrival of the specimen. When culture is prolonged, interim reports may be issued to the clinician. Specimens may also be used for research purposes.

When tests are completed results are appended to the original request and a final report is compiled, verified and issued. Significant results such as a positive blood culture may need urgent action. Details of many infections are reported to the PHLS Communicable Disease Surveillance Centre (CDSC), Colindale to contribute to national infectious disease statistics.

In a laboratory without computers data handling is of necessity undertaken manually. The four most important steps are:

(i) entry of patient (specimen) registration data into a daybook;
(ii) recording of test results on to request forms;
(iii) issue of test results to users;
(iv) filing of test results.

In understaffed laboratories steps (i) and/or (iv) are sometimes omitted, leading to a lower quality of service. Without a daybook the laboratory may not be able to acknowledge receipt of specimens and without access to filed reports the interpretation of current results on a particular patient is less satisfactory. The fifth step—analysis of results to obtain management and infection control information is often not possible at all.

2.2 **Workload measurement**

Measurement of the workload of the laboratory is needed for internal management purposes and may be required by hospital administrators. The best way to measure workload is the subject of hot debate. Specimen numbers may be misleading; relative workload weightings for different specimen types (e.g. Welsh or Canadian units) may not be accurate and assessments based on sampling are open to statistical error. Workload statistics are nevertheless extremely important as they are used to argue the case for resources.

2.3 **Korner data**

As part of a complete review of management information systems in the NHS, a committee chaired by Mrs Edith Korner, Chairman of the Exeter DHA, recommended to the Department of Health and Social Security that to assess the level of activity in pathology laboratories a minimum data set for each request should be recorded. This data set consists of:

(i) the category of the patient (whether NHS or private);
(ii) clinical team reponsible for initiating the request;
(iii) whether the request is for a patient in another district;
(iv) if the request is received out of hours.

The committee also established a definition (albeit contentious) of a 'request'. The Korner recommendations have been accepted and collection of data was due to begin in April 1987. A microbiology laboratory already equipped with a full computer system or an 'electronic daybook' can collect such information without difficulty. For laboratories which have abandoned the use of a daybook, either paper or electronic, the most economical solution may be a microcomputer (or a small network), a database package and sufficient clerical support. As DHAs have been given no specific funds to implement the Korner recommendations it remains to be seen whether these data will actually be collected.

2.4 **Epidemiological analysis**

As well as providing a routine diagnostic service most clinical microbiology laboratories keep records of infectious disease statistics for epidemiological purposes. This data is needed both locally and nationally. Unless the prevalence of organisms and infections within the community is known with reasonable accuracy, a change in trend will not be recognized and the chance to pick up an outbreak at its beginning will be missed. Equally important, the 'background level' of infection will not be mistaken for an outbreak. Continuous monitoring of bacterial isolations, patterns of antibiotic resistance and prevalence of viral infections such as hepatitis B, HIV, measles, rubella and influenza is undertaken to keep local medical practitioners informed and to provide guidance on appropriate antibiotic prescribing.

In addition, most clinical microbiology laboratories in the UK make returns to CDSC or to the Communicable Diseases (Scotland) Unit. At these centres, data is compiled resulting in the issue of weekly reports of infectious diseases and the preparation of longer-term statistics at appropriate intervals. These statistics are used by the DHSS and by Government to provide an up to date picture of infectious diseases in the UK so that national trends can be followed.

The value of central collation of infectious disease statistics was amply demonstrated in an unusual *Salmonella* outbreak in 1982 (5). Increasingly frequent isolations of a rarely seen *Salmonella* (*Salmonella napoli*) in several laboratories in the south-east of England provided early warning of a probable common source outbreak. A case control study was set up quickly by CDSC which established that most of the victims were young and that there was a strong association with consumption of imported chocolate bars. Laboratory testing confirmed the contamination of two particular products which were withdrawn from the market; subsequently 32 tons of contaminated chocolate were destroyed!

National infectious disease statistics, derived from locally collected figures, are also used to monitor the efficacy of current vaccines and to identify the need for new products. An eight-laboratory PHLS survey of rubella immunity undertaken from 1984 to 1987 (6) provided important evidence that the elimination of congenital rubella syndrome was impossible with the current policy of immunization solely of teenage girls and helped to lead to a change in policy to immunization of both boys and girls in the second year of life, with the objective of eliminating endemic rubella from the UK. The monitoring of infections due to human immunodeficiency viruses (HIV) is of paramount importance to government and populace alike. Computerization, if properly planned and undertaken, provides the means to record and rapidly analyse such data permitting management decisions locally or at governmental level to be taken on the basis of the most recent and accurate information.

3. TYPES OF COMPUTER SYSTEMS

Surveys within the PHLS and informal discussions with a number of NHS microbiology laboratories show that there are three principal levels of complexity in the use of microprocessors in clinical microbiology laboratories.

3.1 **Instrument control and direct data capture**

The simplest application of this type is the dedicated mass-produced microchip running items of equipment such as diluters, zone readers, enzyme-linked immunosorbent assay (ELISA) readers, automated blood culture machines, etc. The electronic circuitry is an integral part of the structure of the instrument and cannot be adjusted other than by simple keyed instructions by the user, that is the complexity is broadly similar to that of a pocket calculator.

Microcomputers may be used to capture and process data from instruments. Commercial examples include automated bacterial identification and antibiotic sensitivity testing systems in which analogue signals are converted into test results and linked with particular patients by the microcomputer. 8-bit microcomputers such as the BBC Model B in the UK or the Apple IIe in the USA have been used extensively for various data capture applications developed 'in-house'. Scope for such systems will undoubtedly increase hand in hand with laboratory automation and such devices will in future need to pass results directly into larger laboratory computer systems capable of more sophisticated data processing.

3.2 **Microcomputers**

3.2.1 *Word processing*

The paperless office has yet to arrive. Word processing is by far the most popular application for microcomputer software. A recent survey of PHLS laboratories showed that almost all use at least one microcomputer word processor. The reasons are not hard to find. The software is now thoroughly reliable and generally user-friendly. There are very wide applications but word processing is particularly valuable in three areas. The first is in the upkeep of documents which remain broadly similar over long periods of time but which need intermittent minor amendment. Some examples include:

(i) general practitioner telephone number lists, which need updating to take account of retiring partners and changing trainees but which otherwise remain the same from year to year;
(ii) lists of laboratory staff with home telephone numbers;
(iii) method sheets for the various laboratory benches;
(iv) control of infection, antibiotic and disinfectant policies;
(v) cumulative lists of identified high risk patients such as those known to be hepatitis B or HIV positive.

All these are examples of documents which can be updated and reprinted as necessary with minimal effort.

The second area of application is in the writing of papers and articles, minutes of meetings, etc., where the facility to correct and to rearrange text is invaluable. Third, word processors are ideal for producing labels (*Figure 4*) and maintaining mailing lists.

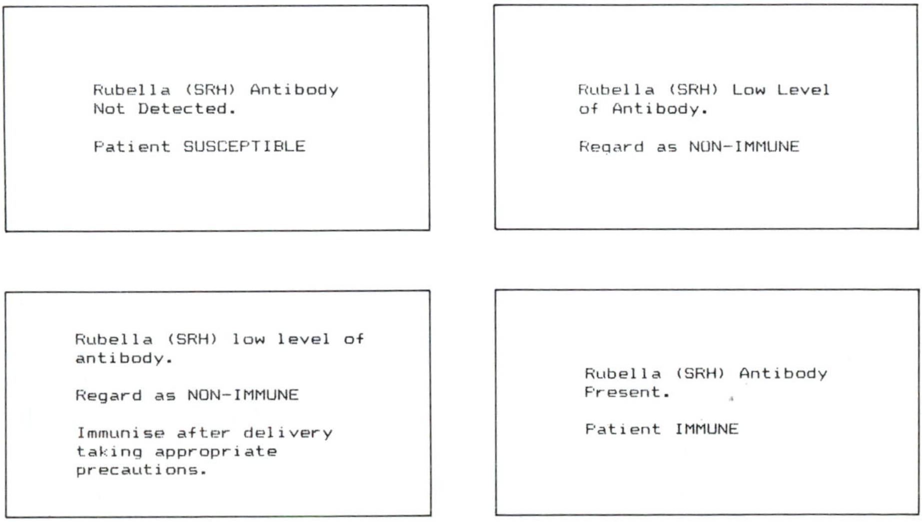

```
Rubella (SRH) Antibody
Not Detected.

Patient SUSCEPTIBLE
```

```
Rubella (SRH) Low Level
of Antibody.

Regard as NON-IMMUNE
```

```
Rubella (SRH) low level of
antibody.

Regard as NON-IMMUNE

Immunise after delivery
taking appropriate
precautions.
```

```
Rubella (SRH) Antibody
Present.

Patient IMMUNE
```

Figure 4. Word processor-generated labels. These are used instead of typed reports.

```
GLOUCESTER PHL - RUBELLA SYSTEM -   AMEND PATIENT DETAILS/ADD RESULTS


Surname          :GALSWORTHY
Forename         :PHILIPPA
Lab. number      : 10002
Date             :261184
Address          :24 LONDON RD, GLOS.
Date of Birth    :201058
Surgery          :LON.  97, London Road, Glos.
Hosp/Ward        :
I/S/D/C          :IMM   Immune

        Press C to change the above, R to enter result
    + to move one record forward, - to move one record back
          or [SPACE] to select another record
```

Figure 5. Patient record (fictitious) from rubella serology system using dBASE II. Once known to be immune, further blood samples submitted from this patient would not be re-tested for rubella antibody.

Word processing is highly cost-effective and should be available in every clinical microbiology laboratory.

3.2.2 *Databases*

Applications of databases running on microcomputers are, if anything, more diverse than the use of word processing. There are numerous commercial database packages on the market, all designed primarily for the businessman. They will mostly adapt well to use in the microbiology laboratory and the choice of package hinges on characteristics such as ease of use, the operating system required, the size and maximum number of records stored, networking capabilities, etc.

Relational databases such as dBASE II or III, Delta, Omnis, Condor, Datamaster and Aspect (see Appendix for addresses of vendors) have been used by PHLS laboratories for a variety of different purposes. Some examples are:

(i) the storage and analysis of test results for rubella, hepatitis and HIV serology;
(ii) viral complement fixation tests;
(iii) leptospira isolation and serology;
(iv) antibiotic sensitivity test data;
(v) the generation of statistics for the production of a regional communicable disease newsletter.

In the Gloucester Public Health Laboratory rubella serology results have been stored on microcomputers for 5 years (*Figure 5*). The initial objective was to identify those women who had been found to be immune in a previous pregnancy thus saving re-testing. At first a database package developed 'in-house' running under CP/M on a Z80 64 K micro was used but this was replaced with dBASE II running under MS-DOS on an Apricot Xi when the floppy disc drives of the original system proved unreliable. Transfer of the data from the old software to dBASE II was achieved without too much difficulty as the data could be unloaded from the first system in ASCII format.

It was found that dBASE II required a reasonable level of programming skill to implement even small changes in input screen formats or in the types of analysis used and the version which we used (2.41) proved troublesome from time to time due to bugs in the indexing. dBASE II is not only a database but is really a programming language as well. When we needed to increase greatly the number of fields of information stored and to change the methods of analysis in order to participate in the multi-laboratory PHLS rubella survey described earlier, a major re-write of the programs proved necessary. Though all the other laboratories participating in the rubella survey were using a different database (Delta) it was ultimately easy to merge data (even that from Gloucester) and to present results graphically (using Deltagraph).

Of the databases mentioned above Delta is perhaps the easiest for the unfamiliar user to develop to his own application. Input screen formats are easily designed, file handling is satisfactory but flexilibity of analysis is perhaps more limited than some of the other packages. Delta has been used by the Manchester PHL to establish a hepatitis B serology database in which a variety of markers of infection are recorded. Data are analysed partly by the facilities contained within the package itself and partly by unloading the data so that BASIC programs can be used. Epidemiological information thus derived is distributed to interested laboratories in a regional communicable disease newsletter.

3.2.3 *Databases in research*

Another ideal use for a microcomputer database is to store and manipulate data from a research project of finite size. Recently we have used such a system to analyse data and results from over 6000 subjects enrolled in a survey of meningococcal carriage and immunity in Stonehouse, Gloucestershire in November 1986 (7). The entire population of the town, together with all staff and children at Stonehouse schools, were invited to provide a throat swab, blood and saliva, together with simple personal details. Seventeen items of data were recorded for each subject (name, date of birth, etc.). Participants completed a registration card and then data were transferred to two Apricot Xi microcomputers running under MS-DOS and using the Aspect database package. The 'slave' microcomputer records were added to the 'master' file each morning throughout the 2-week survey period using floppy discs, taking care that apparent duplicate registration numbers were rejected; the 'slave' datafile was then deleted daily. Two Public Health Laboratories and a university bacteriology department were involved in the testing of the specimens and in data manipulation. All used Apricot microcomputers with the same software. Results of the different groups of tests—meningococcal groups, types and antibiotic sensitivity test results, serology, and blood groups and immunoglobulin levels—were stored in subsidiary records linked to the patient 'master' record by a unique number. Updated results files were sent between laboratories by floppy disc. Compatibility of software and hardware made the transfer of data between laboratories a simple matter. Aspect is one of the few database packages available which will work in a microcomputer network. It is also relatively friendly, particularly in permitting first-time users to perform sophisticated analyses of data (*Figure 6*).

3.3 **Total laboratory systems**

3.3.1 *Types of systems*

Only a minority of UK clinical microbiology laboratories have been equipped with a total laboratory computer system as yet. Several mainframe or minicomputer systems have been marketed, of which the most widely known are perhaps the Ferranti Phoenix

a

Outbreak strain carriers by area and by date of birth.

Area	Date of birth	Lab. No.	Sex	Group	Type	Sulpha.
1	03/07/49	2780	F	B	15/P1.16	R
	17/10/52	2719	M	NG	15/P1.16	R
	19/05/53	3919	F	B	15/P1.16	R
	06/11/54	3889	F	B	15/P1.16	R
	12/10/55	2518	F	B	15/P1.16	R
	27/02/59	2606	F	B	15/P1.16	R
	25/04/63	1093	F	NG	15	R
	27/10/65	2771	M	B	15	R
	22/01/67	3838	M	B	15	R
	06/03/67	569	M	NG	15/P1.16	R
	29/03/67	997	M	B	15	R
	29/04/68	1284	F	NG	15/P1.16	R
	29/08/69	1134	F	B	15/P1.16	R
	22/10/70	5314	F	B	15/P1.16	R
	08/12/70	2770	M	NG	15/P1.16	R
	26/08/71	1142	M	NG	15/P1.16	R
	23/05/74	6246	M	NG	15/P1.16	R
	12/03/75	6120	M	B	15/P1.16	R
	28/03/75	6172	F	B	15/P1.16	R
	02/07/75	6195	M	NG	15/P1.16	R
	06/11/76	5058	F	B	15/P1.16	R
	04/06/79	132	M	B	15/P1.16	R
	19/02/81	5441	M	B	15/P1.16	R
2	06/05/46	804	M	B	15/P1.16	R
	15/10/48	1395	M	NG	15/P1.16	R
	14/01/56	1784	F	NG	15/P1.16	R
	29/05/58	676	M	NG	15/P1.16	R
	15/03/60	657	F	NG	15/P1.16	R
	28/09/60	675	F	NG	15/P1.16	R
	07/10/60	1619	M	NG	15/P1.16	R
	19/06/64	1620	M	NG	15/P1.16	R
	27/05/65	975	M	B	15/P1.16	R
	05/08/70	1675	F	B	15/P1.16	R
	17/08/77	5090	F	B	15/P1.16	R
	05/04/78	144	M	NG	15/P1.16	R
	20/07/78	199	M	NG	15/P1.16	R
3	13/04/48	2415	F	NG	15/P1.16	R
	24/05/50	2227	M	B	15	R
	11/03/51	6804	F	NG	15/P1.16	R
	19/06/53	4733	M	NG	15/P1.16	R
	13/01/61	4105	M	NG	15/P1.16	R
	27/03/62	2179	M	NG	15/P1.16	R
	20/09/63	2178	F	NG	15/P1.16	R
	15/05/65	2443	M	B	15/P1.16	R
	16/05/65	4205	M	B	15/P1.16	R
	12/06/66	4550	F	NG	15/P1.16	R
	20/08/69	4478	M	B	15/P1.16	R
	07/10/78	189	F	B	15/P1.16	R
	22/02/80	5228	M	B	15/P1.16	R
	13/08/80	5510	M	B	15	R
	15/10/80	5438	M	B	15/P1.16	R
	02/04/82	5229	F	NG	15/P1.16	R
4	03/11/46	1040	F	B	15/P1.16	R
	10/02/67	971	M	NG	15/P1.16	R
	30/03/73	1371	M	B	15/P1.16	R
5	01/03/57	904	M	NG	15/P1.16	R
	25/06/63	3191	F	NG	15/P1.16	R
	25/12/74	4648	M	B	15/P1.16	R

Figure 6. The Stonehouse Survey—data stored in Aspect. (**a**) Meningococcal outbreak strain carriers sorted by area and by date of birth. Aspect can present data as lists or as counts. Calculations can be performed on numeric values contained within fields. (**b**) Numbers of children of each sex attending Stonehouse schools with the numbers of each sex carrying *N.meningitidis* and *N.lactamica*.

b

Sex ratio of schoolchildren

School	Sex	No.
MA	F	391
	M	345
TOTAL School MA		736
PI	F	93
	M	120
TOTAL School PI		213
PJ	F	116
	M	110
TOTAL School PJ		226
SH	F	22
	M	28
TOTAL School SH		50
TOTAL		1225

Meningococcal carriers

School	Sex	No.
MA	F	56
	M	49
TOTAL School MA		105
PI	F	6
	M	10
TOTAL School PI		16
PJ	F	7
	M	9
TOTAL School PJ		16
SH	F	1
	M	2
TOTAL School SH		3
TOTAL		140

Lactamica carriers

School	Sex	No.
MA	F	14
	M	11
TOTAL School MA		25
PI	F	4
	M	7
TOTAL School PI		11
PJ	F	4
	M	1
TOTAL School PJ		5
SH	F	3
	M	3
TOTAL School SH		6
TOTAL		47

system installed in Wessex Region laboratories, a system developed at University College Hospital based on Digital Equipment Corporation computers and Mumps system (3), and the Information Technology Limited Microlab system installed in five Public Health Laboratories. Other laboratories have developed networked microcomputer systems. All these systems attempt with variable success to manage the flow of data from the arrival of the specimen plus request form in the laboratory, through the production of results and issuing of the completed reports, to the compilation of management data and epidemiological statistics.

Many advantages of such systems are obvious; they include automated filing, sorting and printing of legible reports, analysis of workload for management purposes, including cross-charging, and analysis of results for epidemiological purposes. They may permit a faster response to telephone enquiry about a particular specimen. Indeed, they may enable a large laboratory to confirm that a specimen has in fact been received.

It is possible on such systems to check for details of previous work undertaken on the patient but in most laboratories this facility is not extensively used. Recall of details of previous specimens is of particular importance in virology departments when paired serum specimens arriving in the laboratory 10–14 days apart must be examined for a rising antibody titre to a particular group of viruses or bacteria such as respiratory or neurological pathogens.

Handwritten request forms are almost universal at present and if the laboratory is computerized data entry is almost always by typing with QWERTY keyboard. Much registration data can be entered in code form—the source address, the clinical team or GP originating the request, the nature and site of the specimen and the investigation(s) required. The clinical details may also be partly coded but most systems recognize the need for free text entry as well.

Likewise, on completion of laboratory work, entry of results on to the computer can be speeded up by good software design and by coding. For a batch of work with large numbers of similar results, such as midstream urine specimens yielding no growth or no significant growth on culture, or ante-natal serum specimens containing adequate levels of rubella antibody, that is rubella immune, the quickest method of result entry is in batches. The computer is asked to generate the particular result ('no growth' or 'rubella immune') for a batch of specimens whose registration numbers are then entered as a group. This method must be used with care to ensure that registration numbers are not placed in the wrong group and thereby inadvertently allocated the wrong result.

For results which demand a more detailed report, for example the isolation of *Bacteroides fragilis* from a wound swab together with a range of antibiotic sensitivity test results, coding again saves time. *B.fragilis* can be reduced to a specific three- or four-letter code and entry of this code prompts the computer to generate a screen giving the names of the usual antibiotics tested against this organism so that results as S (sensitive) or R (resistant) can be entered as single keystrokes.

When work on a specimen has been completed and results entered, verification and/or authorization of the report is undertaken. This may be a two-stage operation, with verification of the technical aspects of the work performed by an experienced medical laboratory scientific officer followed by a medical authorization with the opportunity to add interpretive comment and, if necessary, to take immediate action such as a telephone call or ward visit. In practice most laboratories have only a single authorization step, usually undertaken by a medical microbiologist. Some classes of routine results such as negative urine cultures and rubella immune sera may be issued without verification as interpretive comment is not normally needed and the verifier relies upon his technical staff to have performed the tests and recorded the results accurately.

4. EVALUATING COMPUTER SYSTEMS

A total laboratory computer system is likely to be the most expensive equipment purchase made by the laboratory during the professional lifetime of its staff. Running costs can take up a disproportionate amount of the laboratory budget. Smaller systems and stand-alone microcomputers are less daunting but are still capable of the consumption of enormous amounts of resource. How then are computers for microbiology laboratories best evaluated? They should be regarded as pieces of laboratory equipment like any other, which must be justified before the capital and ongoing costs are incurred. In the past some laboratories have purchased a computer system first and have then applied it to a problem area. This is the wrong way round (though understandable given the current method of NHS budgeting with 'windfall' capital sometimes becoming available at the end of a financial year). It is most important to start by defining the problem and considering whether a computer solution is appropriate. Many tasks are more economically and efficiently performed using pen and paper, simple filing systems, etc. An example is the use of an 'Acute Patients' file in a small or medium-sized laboratory. In our laboratory (specimen numbers ~115 000 per annum with two medical microbiologists) there are normally 20–30 patients at any given time whose cases are of particular concern. Despite enthusiasm for the application of computing we find it simpler and quicker to keep a paper card record of such patients constantly updated than to establish a computer card index to hold the information. Paper cards can conveniently be carried round the wards and are amenable to some degree of simple analysis at the year's end. For larger laboratories, an electronic database (8) or word processor may be more effective solutions but the principle is an important one. Not everything needs computerization.

Having identified a particular problem which can be eased or solved by use of a computer, the next step is to investigate software. There may be particular conditions to be met (size of records to be stored, maximum field lengths, speed of analysis, etc.) which may limit the range of possible software. For more exotic applications software may need to be specially designed and written but, if so, particular care should be paid to its future maintenance and upgrading.

For many microbiological computing applications a wide range of commercial software and hardware will serve. It is advisable, even for small systems to use a checklist (*Figure 7*) for both hardware and software to improve the chances of a successful installation.

4.1 **The manufacturer**

If support will be needed in future it is worth buying from a large, financially secure manufacturer. This does not need to be IBM or Ashton-Tate but caution is called for when contemplating a purchase from a very small or very young company. The dangers of complex systems developed in-house and dependent upon a single person for support and upgrading are obvious. Such systems are rarely transportable and, in the absence

```
Financial health of manufacturer

Dealer support - maintenance, insurance

Compatibility with local users

Sizing - processor power

         data storage - on-line & archive

Siting of system - optimal hardware environment

                   ease of use

                   security

Training of users

Potential for upgrading - software

                        - hardware

Potential for unloading data

External links

Contingency plans for total system failure
```

Figure 7. Checklist for evaluating new computer systems.

of the originator, it may be impossible to obtain help when problems arise. The data is the most valuable part of any computer system. It must be made secure against accidental losses including failures of hardware and software.

4.2 **Costing**

Costs are broadly divided into two groups—capital and recurrent (*Figure 8*). Many computer systems purchased in the NHS today are required to finance themselves by revenue savings. Whether they do so depends upon one's view of 'creative' accounting. The real costs are high. The lifespan of a minicomputer may be about 7 years and microcomputers, especially if hardworked, may last no longer than 3–5 years. The capital cost of the installation should therefore be written off over this period. Maintenance contracts are often about 10–12% of the capital cost and may be higher for an application which is time-critical, that is where computer breakdown even for a short period cannot be tolerated. Software (and its users) will inevitably evolve and allowance for upgradings and changing user demands must be made. By comparison consumables are cheap but nevertheless should be costed accurately. Installation costs should take into account any ancillary equipment required. This will include mundane but easily forgotten items such as building alterations, air conditioning, tables and chairs, in addition to the more obvious accessories such as unusual input or output devices, links to other computers by cable or by telephone, 'slave', VDUs, anti-glare screens etc.

A. Capital

i) hardware - computer (CPU, terminals etc.)
peripherals (printers, input devices etc.)

ii) software - including customising

iii) installation - air conditioning, new building
wiring, security
links to other systems

iv) ancillary equipment - telephones, tables, chairs etc.

B. Recurrent

i) Hardware - maintenance & upgrading

ii) Software - maintenance & upgrading

iii) Staff - new staff
regradings

iv) Consumables - stationery, power etc.

Figure 8. Costs of a computer system.

4.3 Staff

Insufficient attention is given to the changing pattern of work which may result from the introduction of a computer system. The proportions of technical, secretarial and ancillary staff may need altering. The type of work carried out by individuals may alter sufficiently to necessitate a formal review of job description with, perhaps, a change to a new pay scale. Additional staff may be required either temporarily or permanently.

4.4 Compatibility

When there is a choice of computers capable of running the chosen software the next most important consideration is compatibility with other potential users of the data. Conforming to the district or regional hardware policy may bring useful advantages including support of equipment. Hardware policy in the remainder of the Pathology Department and in the hospital must also be considered. Laboratories which share work would find it convenient to use compatible systems. The PHLS has operated with some success a hardware policy to facilitate both equipment support and interlaboratory communication.

4.5 Data transfer

Compatibility is perhaps not quite so critical now as it was only a few years ago. With effort it is now possible to transfer data between almost any two computers. Software should be capable of unloading data files into ASCII format, that is files should be capable of being TYPEd on to the screen. For hardware compatibility a physical means of data transfer is necessary. This can be achieved most conveniently using floppy discs with a common format. The newer 3.5-inch floppies are much more robust than their 5.25-inch predecessors and are safer when sent by post. Alternatives are to link two machines directly by cables, or to transfer data by telephone using a modem link (modulator/demodulator) or by use of magnetic tape.

4.6 Siting of a computer

The siting of the components of a computer system needs some forethought and possibly expert advice to get the best out of both the machines and the operators. The CPU of a minicomputer will need to be in an air-conditioned environment. In fact more stress is placed on any computer hardware which has to work in high temperatures, high humidity or smoky atmospheres.

Contrast on VDUs is poor if they are sited in direct sunlight, but anti-glare screens help if a VDU has to be positioned where there is high reflection from the screen. Accidental inoculation of a keyboard with pathogenic organisms does not seem to have been reported yet but common sense suggests some care in the use of computers in areas where microbial manipulations are carried out. They should not be used in close proximity to areas where high risk pathogens are handled as no satisfactory decontamination procedures have yet been devised.

Staff become rapidly fatigued if working at a VDU sited at the wrong height (standard laboratory benches are too high) or in an area subject to constant traffic flow. Printers can be very intrusive particularly if performing large batch tasks. A dot matrix printer working with a stand alone word processor may be acceptable in an open office but a printer which is used for the output of all laboratory reports as a once- or twice-daily batch job will need to be sited in a separate (preferably soundproof) room unless it utilizes one of the silent technologies such as ink jet or laser.

4.7 Sizing

Many computer systems in the NHS seem to be installed undersized. This is not necessarily because of bad planning but often because of capital cost constraints. Such constraints are usually later perceived as false economies, leading to frustration of the users and either premature abandonment of the system or an early upgrade.

There are two aspects to consider when sizing. The first is the amount of computing power needed. This is visible to the user as the speed of response at VDUs and has been a particular problem with minicomputer systems, though there is ample scope for the same problem with networked micros. Either type of system may be designed to support a particular number of terminals on the basis that not all the terminals are in use simultaneously. In a total microbiology system there are times of the day when most or all terminals will be in use together. Mid-morning is often the peak period and delays at VDUs in a badly undersized minicomputer system may run to several seconds—enough to cause intense user frustration. Networked systems slow down as a result of heavy use of the net and the file server. This is a problem of software design. At peak periods maximum use should be made of the distributed processing power of a network with as little traffic flow as possible on the net.

The second sizing estimate which needs to be made is of the data storage capacity. The on-line and archive data storage capacity need to be calculated independently. In practice requirements for archive data storage capacity usually greatly exceed initial estimates and allowance should be made for this. Undersized systems can sometimes be improved by data compression (particularly with textual data) or by selective archiving. Undersized on-line data storage capacity means more computer memory, usually only a matter of another board.

Some aspects of microbiological work such as HIV antibody results and ante-natal rubella serology require almost indefinite on-line data storage. Other results are more ephemeral. Negative urine cultures may be needed on-line for no more than a week or two and perhaps need no long-term storage at all. The plummeting cost of data storage and the use of exchangeable hard discs and write once/read many (WORM) drives means that, within reason, as much data as is required may be kept in archive form, though in practice recourse to data more than a year old will be limited. A three-tier system offers some advantages. Selected positive results may be kept on line permanently—these might include rubella, hepatitis and HIV serology, positive Tb and perhaps blood culture results; other classes of specimen such as wound swabs, both positive and negative, may be kept on line for a finite period (say 6–24 months) prior to archiving; finally, all results whether positive or negative are held on-line for a short period (4–12 weeks). Many records in this last group may not need archiving at all. Good design, with the facility for flexibility and selectivity in archiving and an allowance for increasing workload, should permit a correct and realistic sizing of both on-line and archive data storage requirements.

4.8 **External links**

Communications between computers have improved rapidly in recent years, and any computer terminal can now act as the key to a diverse range of information and facilities mostly accessed via the telephone system. Prestel and Oracle are well known in the UK but perhaps of more use to microbiologists is the opportunity to perform a literature search, to use Excerpta Medica, to obtain listings of strains held by national microbiological culture collections (MiCIS) or to make use of the Joint Academic Network (JANET). Electronic mail services are also available; these are expensive but offer exciting possibilities for linking users to the laboratory. Such users might include laboratory staff on call, family practitioners, other microbiology laboratories and the PHLS Communicable Disease Surveillance Centre.

4.9 **Training of users**

Until recently vendors of computer systems whether large or small made scant effort to introduce the purchaser to his new technically formidable acquisition. Opaque 'instruction' manuals thicker than a telephone directory did nothing to allay fears, and promises of telephone support were all too often left unfulfilled. The onus remains with the purchaser to ensure that any computer system is delivered with sufficient training to enable the users to get off to a good start. The minimum is a good manual of instructions but anything more substantial than a small software package may require on-site training or secondment of staff to a training course. Telephone support is invaluable. More can be achieved in 5 min over the phone than in an hour with a manual.

If a system is genuinely user-friendly its installation and management should not demand specific programming skills on the part of the laboratory staff. Unfortunately, for large systems in particular, keeping the system running may demand a substantial investment in time and the acquisition of new skills by more than one interested member of staff. This 'hidden cost' of computerization is not sufficiently appreciated and causes problems when a demanding computer system is installed in a laboratory where staffing levels are already marginal.

4.10 **'Future-proofing'**

There is of course no such thing! Nevertheless when buying a new system it is sensible to ensure maximum flexibility provided that this is not at the expense of speed of operation. Over the normal lifespan of any computer system it is certain that there will be technical improvements in the hardware and also in branded software. Equally certainly the purchasers will wish to modify the way in which the system is used leading to a constant evolution of ideas. It is almost unheard of for a computer system to remain unchanged throughout its working life. Useful questions to ask include the following.

(i) How will the data be unloaded when the system reaches the end of its useful life?
(ii) Can the data be transferred to industry-standard hardware?
(iii) Can the data be analysed flexibly in ways which are not yet envisaged?
(iv) Can hardware be upgraded to improve processing speed, memory size or external communication?

5. MANAGING A COMPUTER SYSTEM

The physical environment necessary for computers has been mentioned briefly. Providing a dust-free, cool environment will prolong the life of electronic devices in addition to reducing the risk of breakdown.

5.1 **Security**

5.1.1 *Hardware*

Government departments in the UK do not insure against loss of equipment whether by fire, theft, flood or any other cause. Physical security of hardware therefore needs to be taken seriously. Equipment should, if necessary, be locked up when not in use and protected by appropriate security arrangements.

5.1.2 *Software*

The same strictures apply to software. Vendors of commercial packages request licensed users to keep software secure from copying and many packages are copy-protected. Such protection is only comparative.

5.1.3 *Data*

More important by far than security of hardware and software is the protection of data because this is unique to the user and is usually irreplaceable. Confidentiality of sensitive data is considered below in the section on the Data Protection Act. Equally important is the taking of appropriate measures to safeguard against the loss of data by accident, theft, fire or other mishap. It cannot be over-emphasized that back-up and safe storage of data is the most important single task in the management of a computer system. Back-up of data files must be performed sufficiently frequently to ensure that a catastrophic loss of data *cannot* occur even if a computer is stolen, destroyed or breaks down irretrievably. Most electronic data in clinical microbiology laboratories is held on hard or floppy discs and is simple to back-up, tape streamers or floppy discs providing an economic method. For less important data a single back-up may be sufficient but for all irreplaceable data there should be no less than two back-up copies kept regularly updated. A log should be maintained of the dates and times when back-up copies are made. Back-up copies of particularly important data should be kept in different physical locations so that data is secure against fire, that is in a different building. Whilst there are companies specializing in the recovery of data following disc 'crashes' such recovery is not guaranteed and recourse to such measures can be avoided by correct back-up procedures.

5.2 **Data Protection Act**

The Act imposes a number of duties on the managers of computer systems holding personal data to ensure that such data is:

(i) obtained and processed lawfully;
(ii) held only for specified purposes;
(iii) not used or disclosed other than for the specified purposes;
(iv) adequate but not excessive for the specified purpose;
(v) accurate and up to date;
(vi) not kept for longer than necessary;
(vii) kept secure against loss or unauthorized disclosure.

The Act also permits individuals right of access to their personal data.

In the UK systems holding personal data, which includes anything from personnel records to HIV antibody test results, must be registered with the Data Protection Registrar. Failure to register a system is an offence under the Act. Application forms for registrations can be obtained from Post Offices or from the Office of the Data Protection Registrar (see Appendix for address). It is important to remember that each new system needs separate registration. Systems holding personal data should be restricted to necessary users and password-protected. Printing of data must be limited to what is needed and such printed data must be destroyed after use. Clinical microbiology laboratories hold much sensitive data and the Act places an obligation on the users of such data to ensure that it is managed properly.

5.3 **Daily management**

Regular management is needed particularly for total laboratory computer systems but also to a lesser extent for smaller systems and microcomputers. Back-up of data is the most important priority, but there may be batch jobs to run, maintenance and servicing to arrange, software alterations or enhancements, investigation of hardware or software faults and training of staff. All this is time-consuming. Most minicomputer microbiology systems have required the designation of a system manager who spends 2–4 h on average each day on these tasks.

6. FUTURE DEVELOPMENTS

Much as successive advances in virological techniques have led to the discovery and characterization of new viruses and advances in genetics and molecular biology have led to the development of genetic engineering, so it seems likely that advances in computer technology will lead to radically new approaches to data handling problems in laboratories.

Predictable developments in the near future include the following.

(i) Further increases in the power and speed of operation of computers; there is as yet no sign of slackening in the pace of improvement.

(ii) An increase in the volume of data which may be stored on-line and a further fall in the price per megabyte of long-term data storage.

(iii) Better communications (including networking) between computers due to advances in both hardware and software design.

In the medium term more fundamental changes such as voice and handwriting recognition may revolutionize data entry. The development of so-called 'expert systems' could permit greater automation of laboratory techniques. The effects of these changes will be modified by the way in which advances are used by systems designers. Many of today's microbiology systems seem cumbersome because of limitations in the system design or the software rather than through any hardware deficiency.

The bottleneck in laboratory computing is currently the entry of patient registration data into the system. If hospitals were to install computer terminals on all wards linked directly or indirectly to pathology laboratories, requests could be entered at a keyboard by the users, avoiding duplication of this task. Requests could be sent immediately or could be held in a file to be 'dumped' to the laboratory in a batch, perhaps three or four times each day. The latter approach would obviate the usual complaint about such a proposal—that adding further links to a fully-employed hospital computer system would slow screen response times to an unacceptable level. A direct link between wards and laboratory would also permit reports to be sent by this route and allow the users to interrogate an authorized file of results. In time each patient's notes will be held electronically and not as a paper file, allowing the laboratory to obtain relevant clinical information and up to date results of other tests including radiology and chemistry. This should greatly improve the quality of the microbiology service.

Links between the laboratory and GP users are now technically feasible but expensive. They would bring to the GP a faster service and to the laboratory a welcome reduction in telephone enquiries. Collection of national infectious disease data by electronic mail is also technically within our grasp but would need financing. The flow of information would again be two-way, improving speed and accuracy of reporting.

Systems which rely on a single central processing unit are inherently vulnerable. Greater use of distributed data processing seems likely as such systems offer a number of design advantages, in particular the lesser potential for total system failure and a reduction in maintenance costs. Their current drawback is that they tend to operate slowly and the problems of data flow within the net are only now being satisfactorily resolved.

Those laboratories which have already installed total computer systems or even those which make extensive use of microcomputers know that changes in the work pattern of staff follow inevitably. With current systems every user needs at least simple typing skills. It is a poor use of resource for technical, scientific or medical staff to spend long hours at keyboards typing at a pace slower than that of a secretarial neophyte. Perhaps the greatest changes wrought on our laboratories by computerization will be in work patterns, in the types of staff employed and in the duties which they undertake. It is up to all those working in clinical microbiology laboratories to grasp the opportunities offered by computers to provide an improved service for the users and a more efficient working environment within the laboratory.

7. REFERENCES

1. Ayliffe,P.F. and Chalke,R. (1973) *Med. Lab. Technol.*, **30**, 363.
2. Mitchison,D.A., Darrell,J.H. and Mitchison,R. (1978) *J. Clin. Pathol.*, **31**, 673.
3. Ridgway,G.L., Batchelor,J., Luton,A. and Barnicoat,M. (1980) *J. Clin. Pathol.*, **33**, 744.
4. Williams,K.N., Davidson,J.M.F., Lynn,R., Rice,E. and Phillips,I. (1978) *J. Clin. Pathol.*, **31**, 1193.
5. Gill,O.N., Sockett,P.N., Bartlett,C.L.R., Vaile,M.S.B., Rowe,B., Gilbert,R.J., Dulake,C., Murrell,H.C. and Salmaso,S. (1983) *Lancet*, **i**, 575.
6. Miller,C.L., Miller,E. and Waight,P.A. (1987) *Br. Med. J.*, **294**, 1277.
7. Stuart,J.M., Cartwright,K.A.V., Jones,D.M., Noah,N.D., Wall,R.J., Blackwell,C.C., Jephcott,A.E. and Ferguson,I.R. (1987) *Epidemiol. Infect.*, **99**, 579.
8. Gaunt,P.N. (1987) In *Computers in Microbiology*, Rand,J.D., Feltham,R.K.A. and Shepherd,W. (eds), University of Leicester.

8. APPENDIX

Names and addresses of database producers

Databases	*UK*	*USA*
Aspect	Microcroft Technology Ltd The Old Powerhouse Kew Gardens Station Kew, Surrey TW9 3PS Tel. 01-948 8255	UK only
Condor	Condor Software Ltd 2 Alice Owne Tech. Centre 251 Doswell Road London Tel. 01-278 2377	UK only
Datamaster	Sapphire International 180 Cranbrook Road Ilford, Essex Tel. 01-554 0582	Software Solutions POB 1116 Bridgeport CT 06601
Delta	Compsoft Ltd Compsoft Manor Francombe Hill Godalming, Surrey Tel. 048 68 25925	UK only

dBASE II and III	Ashton-Tate (UK) Ltd Oaklands 1 Bath Road Maidenhead SL6 4UH Tel. 0628 33123	Ashton-Tate Ltd 20101 Hamilton Ave. Torrance CA 90505
Omnis	Blyth Software Ltd Mitford House Benhall, Saxmundham Suffolk IP17 1JS Tel. 0728 3011	Blyth Software Inc. Century Plaze 1065 East Hillside Boulevard Foster City CA 90505
Data Protection Registrar		
	Springfield House Water Lane Wilmslow Cheshire SK9 5AX Tel. 0625 535777	

CHAPTER 8

Computers in fermentation

MELVYN C.WHITESIDE and PHILIP MORGAN

1. INTRODUCTION

Computers have been used to control fermentation processes since the late 1960s. The industrial sector was very quick to exploit the potential of computer control, for example the Finnish company Rintekno have been supplying minicomputer-based systems for controlling multifermenter rigs since 1972. However, the high cost of computers precluded the widespread adoption of computer control for small-scale laboratory fermenters until relatively recently. The advent of microcomputers in the late 1970s and their subsequent evolution has meant that computer systems have become very much cheaper, more powerful and more reliable. Consequently, there now exists a multitude of commercial, microcomputer-driven fermenter control systems. In parallel, there have been significant developments in the controllers, amplifiers and ancillary electronic equipment required for accurate control of fermentation parameters.

Anyone encountering fermenter control systems for the first time will discover a bewildering array of jargon, prices and 'black boxes' of electronics. Above all, advice from sales representatives and their literature may only serve to further confuse the novice. The aim of this chapter is to provide a guide for the intending purchaser and the interested user to the different approaches to computer control of laboratory fermenters and to explain the most frequently encountered jargon. We will consider the hardware required, the way in which control is achieved and the types of software available. The concepts are applicable to fermenters of all sizes but no attempt will be made to consider specifically commercial-scale fermentation systems, both because of their complexity and because this aspect of bioengineering has received detailed attention elsewhere (1–3).

2. TYPICAL SYSTEM COMPONENTS

All fermenter control systems possess a number of basic constituents. The fermenter system itself will include the culture vessel, stirrer, gas supplies, heaters and pumps. The fermenter may be employed in batch or continuous modes. In addition, computers can be linked to more novel culture systems such as turbidostats (4) and gradostats (5). Sensors are required in order to monitor parameters such as pH, temperature and dissolved oxygen. The most commonly controlled and logged parameters in small-scale systems are listed in *Table 1*. Further details of these parameters have been given elsewhere (1,3). Amplifiers are normally required to boost weak signals from sensors so that they may be read by the electronics controlling the system. Interfaces are necessary to convert the signals from the sensors to the binary format used by computers.

Table 1. Fermentation parameters commonly logged and/or controlled by fermenter control software.

Controlled/logged parameter	*Comments*
pH	Control by relays to acid and alkali pumps
Dissolved oxygen	Control by stirrer speed or by gas blending
Temperature	Control by heating element or valve-operated 'cold finger'
Foam	Relay operation of foam paddles or pump for anti-foam
Culture density	Control of dilution rate in turbidostat system
Substrate addition rate	Timed or in turbidostat system
Stirrer speed	Independent or as part of dissolved oxygen control
Effluent gas composition	Via on-stream gas analysers
Vessel substrate concentration	Use of selective electrodes
Viscosity	

An analogue-to-digital converter (ADC) converts the voltages from the sensor amplifiers to digital values. A digital-to-analogue converter (DAC) changes the digital outputs from a computer to voltages that can be used to drive variable speed pumps or alter stirrer speed, for example. Simple on–off values, such as that generated by a foam detector or required to operate a relay, need only be processed through digital interfaces. Relays are required to switch power to pumps, heaters and solenoid actuated valves. Finally, and most importantly, it is necessary to have a computer running appropriate software to read data from the sensors and take the action needed to maintain the desired setpoints. The system may be based on a mini- or microcomputer; standard commercial computers or machines designed specifically for control applications can be employed. The software may be specifically designed for microbiological applications or may be a more general process control package. The former is, of course, far easier to implement but may be less flexible. A large number of individual components must be considered in specifying a fermenter control system and operation of some of the key items of hardware will be discussed in more detail.

2.1 The analogue-to-digital converter

Computers cannot deal directly with the continuously varying outside world and signals from the latter must be approximated to finite digital values. It is the job of the ADC to read a proportional voltage (analogue value) from the sensor and produce a digital representation of it that the computer can understand. Sensors such as electrodes and electrical thermometers generate a small voltage proportional to the parameter they are measuring. For example a temperature probe might generate 1 V at 10°C and 5 V at 50°C. This voltage is no use to the computer as it stands and consequently an ADC must be interposed between the sensor amplifier and the computer in order to convert the voltage into a digital value.

The ADC selected always represents a compromise in terms of accuracy, speed and price. The accuracy or resolution of an ADC is expressed as the number of bits it sends to the computer. The more bits it sends, the better the approximation. For example, an 8-bit converter returns a number in the range of 0–255. Thus it can divide a signal voltage within its range into 256 discrete levels. This corresponds to a maximum accuracy of 100/256, which is 0.4%. A 10-bit converter can generate a number in the range of 0−1023 and would have an accuracy of 100/1024 (0.1%). Similarly, a 12-bit ADC would have an accuracy of 100/4096 (0.025%). With the hypothetical 1−5 V temperature probe mentioned above, 8-, 10- and 12-bit converters would give measurement accuracies of 0.16, 0.04 and 0.01 °C respectively. Obviously the accuracy of the conversion required varies with the parameter to be monitored. For instance, measuring pH over a range of pH 0−14, an 8-bit converter divides the signal into 256 levels, that is into pH steps of 0.055 units, whereas the same converter linked to a dissolved oxygen probe measuring 0−100% results in a step size of 0.4%. If the parameter is only to be logged these accuracies are more than adequate. However, if close control is required then a more accurate ADC is necessary for these parameters. For example, to control pH to within 0.1 of a pH unit a measurement accuracy of greater than 0.05 of a pH unit is necessary and, since 8-bit accuracy can only resolve 0.055 of a unit, a 10-bit converter (or better) is required.

Accuracy can be degraded by mismatching the ranges of the ADC and the sensor amplifier. An 8-bit ADC with an input range of 0−5 V converts 0 V to a digital value of 0 and 5 V to a digital value of 255. An input voltage greater than 5 V will still return 255 and one less than 0 V will return 0. It is also important not to overload the inputs to ADC converters as this will damage the sensitive integrated circuits. Protection against small overloads (typically up to 25 V) is normally within the circuitry of an ADC system but higher input voltages may circumvent the protection. Conversely, if the voltage range of the attached pH probe is 0−2.5 V, half the accuracy will be lost. The output voltage of a sensor may be amplified (see Section 2.5) but this may result in the introduction of electronic noise into the signal with a consequent degeneration in the accuracy of measurement. Whilst this is unavoidable for those probes generating a very weak signal, in order to make best use of the interfaces it is important that, wherever possible, the voltage ranges of the sensors are matched to the ADC input range.

The time taken for an ADC to complete a single conversion varies with the accuracy, the higher the accuracy the slower the conversion. In most fermentation processes the conversion speed is irrelevant as conversion times for a single channel tend to be measured in milliseconds which is significantly faster than any corrective action or biologically induced change. Problems can arise though if the user wishes to control a number of rapidly moving parameters, such as dissolved oxygen, with high precision. Most commercial interfaces will have eight (or multiples of eight) analogue input channels, that is connections for eight input devices. However, inside the interface there is usually only one ADC chip because this is normally the most expensive component. In such a case the signals from each input are channeled to the ADC circuit sequentially by another circuit called a multiplexor. As a result of this, calculation of the minimum time between readings in a control cycle requires consideration not only of the conversion time itself but also of the cycling time and the number of channels that are active. For example, a typical conversion time for a 12-bit ADC is 30 msec, thus to convert eight channels in a cycle takes approximately a quarter of a second and this is before any programming slows the process further. In practice few microcomputer-based control systems have cycle times of less than a second and indeed these are not necessary for most simple fermentation systems. If more rapid monitoring is necessary, faster, more expensive ADC converters are available, for example 12-bit resolution in less than

1 msec can readily be achieved. Rapid cycle times may become important in larger and more complex systems in order to achieve smooth responses to rapid changes in demand.

To summarize, the type of ADC interface required depends largely on three parameters, namely the required resolution, the voltage ranges to be measured and the speed of monitoring.

2.2 **The digital-to-analogue converter**

A DAC works like an ADC in reverse. The DAC converts the digital output from the computer into an equivalent electrical signal or voltage. It is this voltage that controls stirrer motors, servo-driven gas blending valves, and variable speed pumps. Similar considerations to those discussed above for ADC systems apply to the selection of a DAC system for fermenter control. If possible, the output voltage of the DAC and the controlled devices should be matched in order to minimize the need for amplifiers. The accuracy of control is determined by the number of bits used, for example, an 8-bit DAC driving a stirrer motor with a speed range of 0–1000 r.p.m. has a minimum variation of 4 r.p.m. which is perfectly adequate. However, greater accuracy may well be needed for other parameters. The cycle time of a DAC may also limit accuracy, although the cycle time will generally be significantly more rapid than the responses of many fermenter controllers.

2.3 **Digital ports**

The digital ports are used to control switchable (on/off) equipment and monitor on–off values. The port is simply a connector for a number of wires which the computer can set individually to 0 V corresponding to off or 5 V corresponding to on. Note that the output port itself does not supply enough power to drive relays directly and an interface is required to increase the voltage and current sufficiently to drive standard commercial 12 or 24 V relays.

2.4 **Relays**

Two types of relay are commonly used for fermenter control, solid-state and mechanical. Solid-state relays are quiet in operation, inexpensive, but easily damaged. Mechanical relays are noisy, more expensive but very reliable. The choice must depend on application.

2.5 **Conditioning amplifiers**

The signals from sensors may be very weak and non-linear with respect to the measured parameter. For example, a pH probe generates no more than a few millivolts that are inversely proportional to the pH. Furthermore, the output is temperature dependent. The signal must therefore be boosted and corrected by a specialized amplifier or conditioner before being fed into the ADC. When choosing amplifiers it is necessary to check whether they will run from the same power supply as other equipment. If this is not the case, a separate supply will be required and extra electrical work will be necessary to enable them to be powered independently.

3. DATA HANDLING

We have discussed how to get the digitized numerical data into the computer and it is now necessary to consider how the software handles this. Once the data enters the computer the first step is to convert this electronic information back into meaningful values. To achieve this the computer must have a calibration table so that it can calculate the actual value. For example, the software will calculate that the value 123 that has been input on channel 7 is equivalent to a pH of 7.56 and will store and/or display this information. Some software calibration routines involve two measurements of the parameter that is being logged and from this the slope of a straight line calibration plot is calculated for use in the conversion functions. This will only function effectively if the response is linear and it is better if the software constructs a multipoint calibration curve. We will now discuss the ways in which a computer can be used for the handling of fermentation data.

3.1 **Data logging**

Perhaps the simplest role a computer can play in a fermentation process is as a sophisticated multichannel chart recorder. Microcomputers such as the Acorn BBC microcomputer, Apple II and IBM PC are often used to collect data from the pH, temperature, oxygen and other probes. For example, the BBC series machines have four 10-bit analogue ports built-in that can be accessed easily from BASIC allowing a simple low-cost system to be constructed by anyone with a little knowledge of interfacing and BASIC programming. Commercial software packages are available for many popular machines that eliminate the need for programming skills and these may provide data collection, printout and storage on disc for post-run analysis. A complete data logging system linked to a fermenter system will normally consist of instrumentation amplifiers, an interface to convert the analogue voltages into digital values, a microcomputer and appropriate software. The interface hardware was described above. The software should be able to collect data at user-defined intervals, print the calibrated values and units on a printer and save the data on disc for later use. This stored data is normally used in graph plotting routines at the end of the fermenter run. More sophisticated packages will be able to plot the collected data on the screen in digestible 6, 12 or 24 h chunks, whilst data collection goes on in the background. This type of display is referred to as a trend or history plot.

A very important consideration is the amount of storage space required for the data. A simple 'rule of thumb' guide is to assume that each parameter stored will require 5 bytes per observation plus 10 bytes for a time and date stamp every time a group of parameters is stored. For example, reading eight parameters at a rate of one observation per minute uses 50 bytes per min ($8 \times 5 + 10$). Thus a 360-K floppy disc can hold 5 days worth of data and a 10-Mb hard disc can hold 277 days worth of data. However, the disc operating system may constrain the maximum size of a file. Although floppy discs are available in a number of sizes, the latest 3.5-inch industry standard developed by Sony is to be strongly recommended for laboratory use since the discs are enclosed in hard plastic cases with a self-closing door to protect the head slot. The cost of Winchester hard discs in now comparable with floppy drives. For greater flexibility it is probably worth specifying a hard disc system from the beginning. Hard discs for PC systems range from 10 to 115 Mb. A bit advantage of hard discs is their capacity to hold data from numerous runs which greatly simplifies direct data comparison. If a hard disc is used it is absolutely vital to make regular backup copies of the data onto tapes or high capacity floppy discs. The latest advance in data storage technology is the earthy sounding WORM drive (the acronym stands for write once read many times). This is a laser disc technology based on the common compact disc player. Each disc can record 200 Mb of data but cannot be erased, making it ideal for archiving bulky raw data. However, the current cost of a system makes it an expensive luxury.

3.2 **Adding data logging to existing systems**

The feasibility of 'bolting on' a data logging system depends on the type of output sockets, if any, present on the back of the existing instrumentation amplifiers. Most recent dedicated controllers and amplifiers are provided with a 0–5 V output, older systems may have much lower outputs, for example 0–10 mV signals are common. If the outputs are in the 5-V range they will feed straight into an ADC but the lower voltages will need intermediate amplifiers. If amplifiers are needed for more than one input, it may make more sense to replace the instrumentation completely since more accurate and reliable results will be achieved.

4. FUNDAMENTALS OF COMPUTER CONTROL

In this section we intend to discuss how computer software actually controls fermenter parameters to maintain them at given set points. The basic unit of control employed in software is the single control loop. For example, acid or alkali addition is used to control the vessel pH. The computer checks the value coming from the electrode (the input channel) and sends signals for corrective action through an output channel, for example to turn on a pump. This constitutes a control loop. A control loop may have more than one channel in existence as is the case if both acid and alkali addition is employed for pH control. To summarize: a control loop is merely an input channel linked to one or more output channels via a test to see whether action is necessary.

4.1 **Types of control**

4.1.1 *Digital control*

The simplest form of control is a switch, perhaps to turn a pump or heater on and off. These digital outputs may come directly from the computer but more often they are sent from the interface that is acting under the control of the computer. A control system will have a series of digital output connections, typically in multiples of eight, which can trigger mains relays. The control lines themselves send a signal of 5 V for on and 0 V for off. This cannot drive relays directly and an interface is therefore necessary. It may also be necessary to electrically isolate the external device from the computer with transformers or opto-isolators in order to prevent electrical spikes damaging the interface electronics. Some pump motors and solenoid actuated valves need 12- or 24-V power supplies. Although a mains rated relay can, of course, be adjusted to switch lower voltages, a 24-V relay cannot switch mains voltages. It is therefore advisable, if budgets permit, to opt for the former in order to obtain maximum versatility for further developments. When ordering a system it is always essential to check the current rating of the switching relays: remember heaters will draw a lot more current than pump motors.

4.1.2 *Analogue controls*

Some control actions need more than a simple on or off control. For instance, a stirrer used to control dissolved oxygen tension will need to be varied from, say, 200 to 800 r.p.m. The control program has therefore to output a graduated response which instructs the stirrer to rotate at different speeds. The speed of a stirrer is normally controlled by a small analogue signal voltage generated by a DAC converter, for example a 0−5 V output signal from a DAC can be calibrated to correspond to speeds from 0 to 1000 r.p.m.

4.2 **Control algorithms**

A computer program must respond in a sensible fashion if it is to control fermentation parameters with reasonable accuracy. Consequently, the control loops in a software package can be constructed to react in many different ways to the information from sensors. The names of these types of programmed responses (algorithms) are frequently encountered in control literature. Consequently it is necessary to possess a basic understanding of the more common methods. Further details of these and the less commonly used programming techniques are available in more specialized articles (2,3).

4.2.1 *On−off control*

This is the simplest form of control. As the measured value of the controlled parameter moves away from the setpoint, the computer responds by switching on a pump, valve or heater. When the measured value returns to or near the setpoint, the pump, valve or heater is switched off. This type of control is inherently cyclic since the value of the controlled parameter oscillates around the setpoint (*Figure 1a*). There is a dead-band around the setpoint within which no action is taken due to limitations in the accuracy and speed of signal measurement and switching response. A more advanced variant of on−off control involves a dual system whereby low output devices are employed when the measured variable is close to setpoint and high output devices are switched on when the value has deviated further. This is termed multizonal control and allows for more subtle control close to the setpoint which tends to reduce the amplitude of cycling.

4.2.2 *Proportional control*

This type of control differs from on−off control since it provides a graduated response to changes in the controlled variable. The proportional control algorithm sets a specific output level for each measured value of the controlled variable within the controlling range (proportional bandwidth). This is the percentage of the range of expected measured values over which the control output changes from fully off to fully on. The term proportional bandwidth is seldom used, instead its reciprocal value, the proportional gain, is employed. This concept can be exemplified by considering a variable gas valve being used to supply air as a means of controlling oxygen concentration. Let us consider

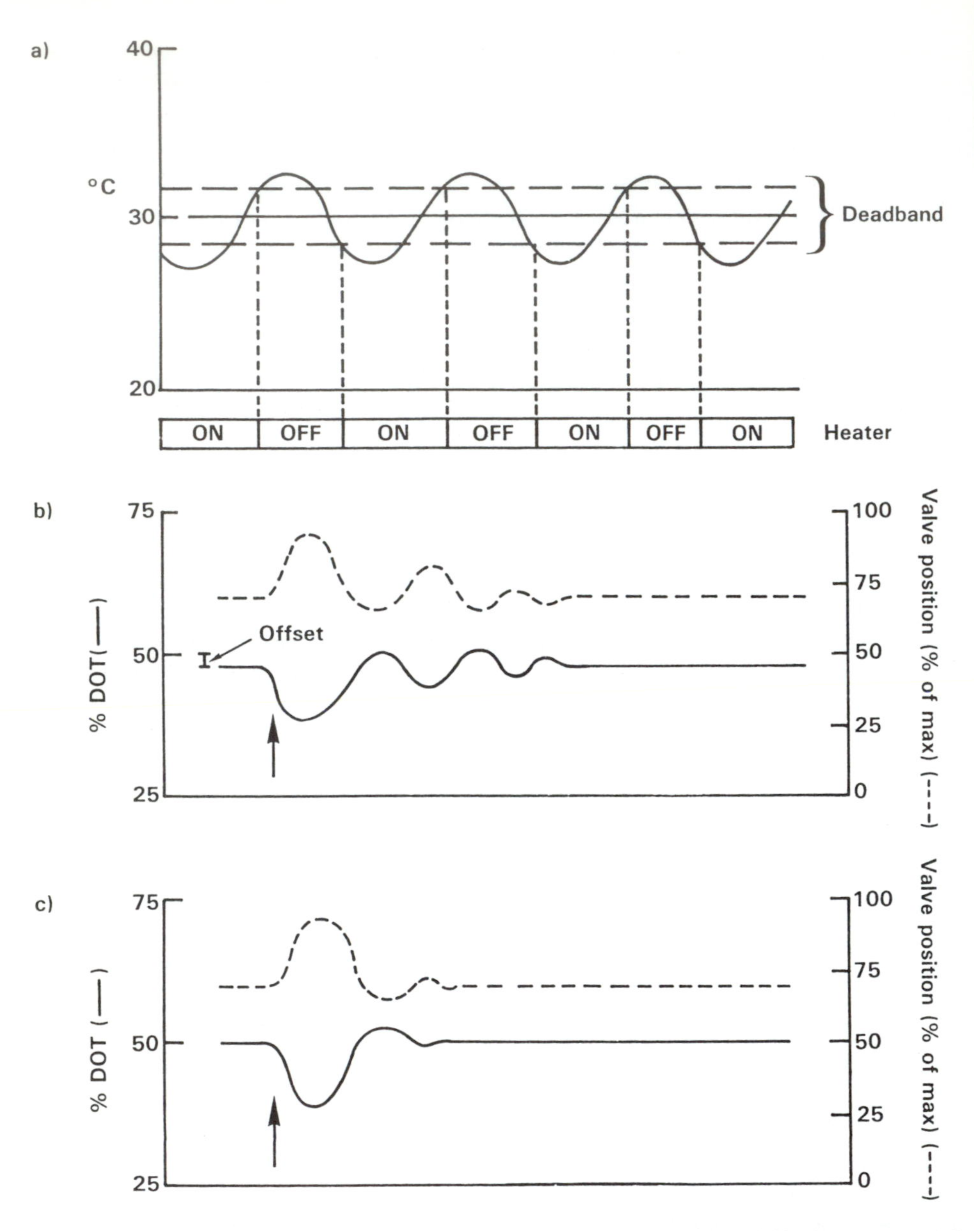

Figure 1. Graphical representations of the three most common types of control provided by fermenter control software. (**a**) Control of temperature by means of a heating element switched by an on–off algorithm. Note the cyclic nature of control due to overshooting and the delays in switching the heater as a result of the deadband. (**b**) Dissolved oxygen control using an air supply valve operated by a proportional algorithm. In the illustration the oxygen uptake has increased (arrowed) causing a drop in the dissolved oxygen concentration below the setpoint of 50%. This is corrected by opening the air valve further (broken line) and thereby increasing the air flow. Note how the control oscillates until a stable value is obtained (solid line) but that this is offset from the setpoint due to limitations in the algorithm. (**c**) Dissolved oxygen concentration controlled using an air supply valve operated by an ideal PID algorithm. The same situation applies as for case (**b**) but note the reduction in oscillation, the faster attainment of setpoint and the absence of offset due to three term control. For full details see text.

a setpoint of 50% dissolved oxygen tension (DOT). The proportional control algorithm could be set with a gain of 2 (i.e. a bandwidth of 50). This means that at 25% DOT the air supply valve is completely open and at 75% DOT the valve is completely closed (*Figure 1b*). The algorithm then controls the DOT concentration by altering the valve position in proportion to the drift from the setpoint. This type of control does possess some inherent inaccuracy. For example, if there was an increase in oxygen uptake by the culture in the fermenter, the dissolved oxygen concentration would fall below the setpoint. The control algorithm would therefore open the air valve to compensate for this and the concentration of dissolved oxygen would rise. However, this would rise to a value slightly below the setpoint due to the phenomenon of offset or droop common to all proportional algorithms. The magnitude of this offset is dependent on the proportional gain setting: a large proportional gain results in a small offset but induces greater cycling around the setpoint. With careful setting of the proportional gain by trial and error, it is possible to obtain smooth non-cyclic control but this will aways be slightly offset from the setpoint.

4.2.3 *PID control*

PID is an acronym for proportional integral derivative control and is also known as three term control. The proportional term has the same function as described above. The other two terms act in parallel with this to overcome the inherent weaknesses of proportional control and thereby increase accuracy (*Figure 1c*). The integral term compensates for the inherent offset exhibited by proportional algorithms by periodically checking for a small deviation from setpoint and adding its own correlation factor to the output device. The number of times per second this integral or re-set action is taken is the integral time. Unfortunately, the integral term also has a de-stabilizing effect on the system in that it increases the time taken to bring the system back to setpoint after a change in load, for example a large increase in oxygen uptake, by adding to the overall input. The derivative term is consequently added to increase stability by countering the effect of the integral term. The derivative action is proportional to the rate of change of the variable rather than the amount of deviation from the setpoint. Thus a rapid change will induce a response from the derivative portion of the algorithm which will damp down the speed of alteration and thereby prevent overshoot.

4.3 **Practical applications of control**

So far we have considered the requirements for assembling a computer system for fermenter control and the way in which the interface hardware and software provide the signals that are required. It is now necessary to discuss the best ways in which some of the main individual fermentation parameters (*Table 1*) can be controlled.

4.3.1 *pH*

Control of pH is achieved by the addition of acid or alkali using peristaltic pumps controlled by relays. Thus when the actual pH in the vessel exceeds the setpoint the acid pump is switched on, conversely when the pH drops below the setpoint the alkali pump is switched on. In its simplest form this can be achieved by on–off control but the technique is unsatisfactory because the pump is on whenever the value deviates from the setpoint. This will result in the value overshooting and thus the pH will oscillate about the setpoint (cf. temperature control illustrated in *Figure 1a*). This situation can be overcome in two ways. Firstly, a deadband or margin can be programmed. This is a small value either side of the setpoint within which no control action is taken. This will prevent excessive addition of acid or alkali close to the setpoint and minimize the risk of overshoot. More accurate control can be obtained by combining the deadband concept with the pumping of less acid or alkali as the set point is approached. This can be achieved by building a proportional term into the control algorithm that controls the length of time for which the pump operates. So if the pH has drifted far from setpoint, the pump is switched on for long periods and as it gets closer the time on will decrease and the time off increase. Therefore with time proportional control of on–off pumps and judicious use of deadbands, accurate control of pH can be achieved using digital outputs and has the added advantage of using the minimum volume of acid and alkali. A more expensive approach is to use variable speed pumps controlled by analogue outputs. pH can then be controlled using standard proportional or PID algorithms.

4.3.2 *Temperature*

The temperature of a fermenter can be accurately controlled by simple switched devices. Heater elements and piped hot or cold water 'fingers' are normally used. The normal technique employed for best results is the use of a heater element in tandem with a valve switched cold water 'finger'. For accuracy, the same techniques of deadband and proportional control as were described for pH apply to temperature control.

4.3.3 *Dissolved oxygen tension*

Control of dissolved oxygen is one of the most difficult tasks in fermentation control. Three approaches are commonly employed.

(i) The concentration of dissolved oxygen in the medium depends upon the flow-rate of oxygen or air passing through the vessel and the mixing efficiency, roughly proportional to the stirrer speed. The simplest form of control is to change the stirrer speed in inverse proportion to the change in DOT. Thus, as oxygen concentration drops, stirrer speed increases and vice versa. This approach works reasonably well as long as foaming is not enhanced and the microorganism in culture has a high demand for oxygen and can stand the shear forces of vigorous stirring.

(ii) For greater accuracy and to control low concentrations of dissolved oxygen, the rate of air or oxygen flow through the vessel must be controlled by a continuously variable valve. This can also be achieved by keeping the flow of gas constant by varying the proportions of nitrogen and oxygen or air in the gas mixture.

(iii) Both stirrer speed and continuously variable gas valves are controlled via the analogue outputs from the computer. Control can also be achieved by combining both methods in a two-tiered system. For example fine control can be maintained using stirrer speed, whilst large shifts in oxygen demand can be accommodated by controlling air flow. The air flow control loop is set with a wide deadband so that it only becomes operative when alterations in stirrer speed alone cannot cope.

4.3.4 *Foam control*

Response to foaming will be triggered by a digital signal from a foam sensor. Foaming may be controlled by means of an anti-foam blade turned on for a user-defined period or by the timed addition of small volumes of an anti-foam solution. The foam control cycle is best re-triggered a number of minutes after the previous corrective action in order to give the foam time to subside after treatment.

4.3.5 *Other parameters*

There are a variety of other fermentation parameters that can readily by controlled (*Table 1*). These will employ the same basic techniques as described above.

5. APPROACHES TO COMPUTER CONTROL

Control of the main fermentation parameters has now been discussed with reference to the hardware required and the way in which control software operates. We have so far described the manner in which the computer itself interacts with converters, relays, amplifiers, etc. There are two fundamental approaches to computer control of fermentation processes. These involve either direct control of the fermenters by the computer software itself or the use of intelligent local controllers that can operate independently of the computer and only communicate with the latter to exchange information. These processes are termed direct digital and supervisory setpoint control, respectively.

5.1 **Direct digital control**

Direct digital control (DDC) is the simpler of the two options. With this type of system a microcomputer and its associated interfaces are directly responsible both for data logging and control functions. The principles of DDC control are illustrated in *Figure 2*. This type of system can be considered as a direct replacement for the traditional control instrumentation and chart recorders. The displays and switches are replaced by the computer screen and keyboard, and the control algorithms are written into the computer software rather than being part of the electronics of the control unit. Calibration of inputs and outputs is software controlled. Additional features are commonly provided by a computer-controlled system which may not be present on older control systems. Examples of these features are discussed in Sections 5.1, 5.2 and 6.2, below.

There are a number of particular advantages to DDC systems. The systems are relatively cheap, highly flexible since aspects of control are accessible through software and can be individually tailored to a process. This is ideal for process development both in laboratory systems and in 'scale-up' experiments. Furthermore, since the control algorithms are written in software, they are relatively easy to modify in response to particular circumstances, unlike electronic controllers where the algorithms are hard wired onto circuit boards. Complex configuration details for the fermenter system can be stored on disc and recalled later for reference purposes and for the instantaneous set up of fermenter runs. Similarly, if the system requires expansion, the addition of extra control loops will not require the purchase of extra hardware as long as there are available channels on the interface because changes can be made in the system set up information that enable these to be used. Further useful features that cannot be

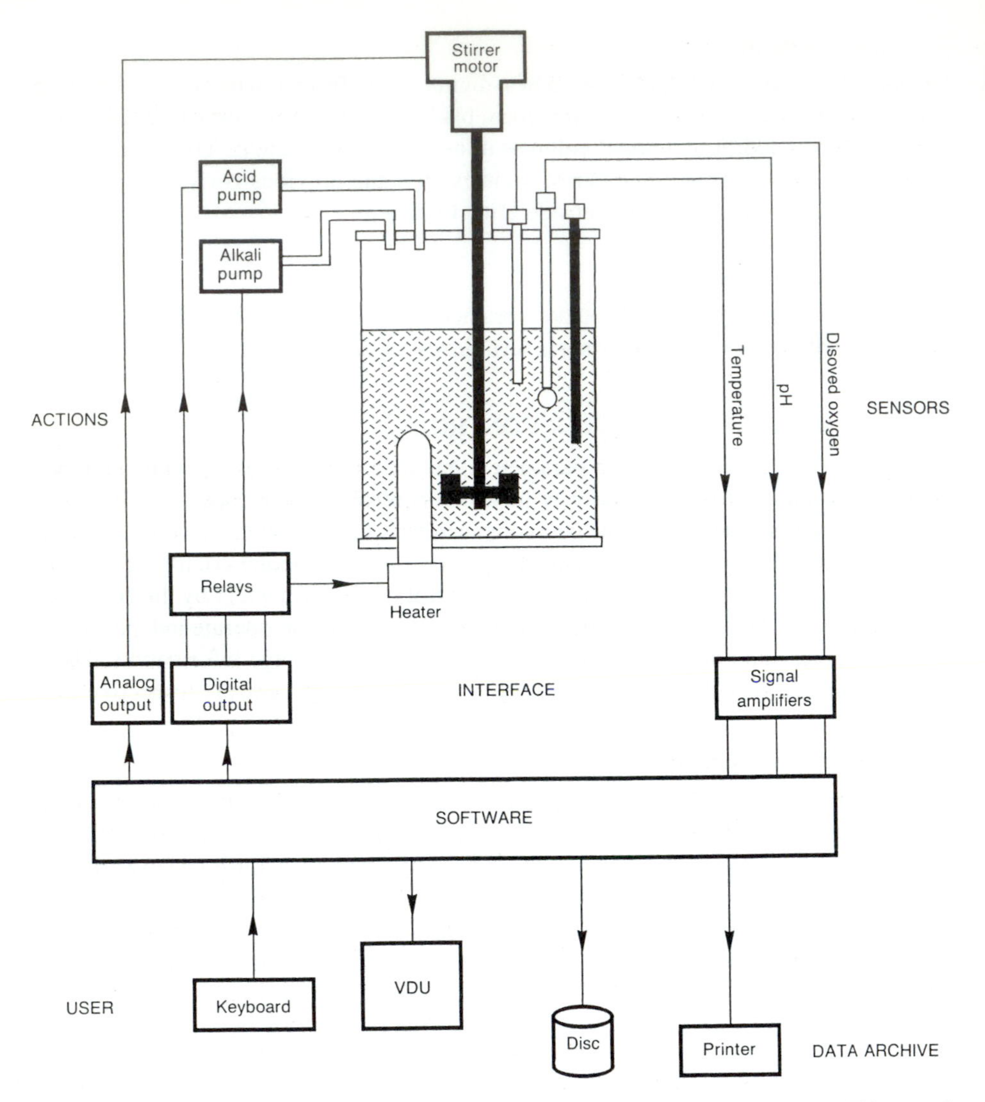

Figure 2. Diagrammatic representation of a direct digital control (DDC) system for fermenters. This example illustrates a system controlling temperature by means of heating only, dissolved oxygen tension by stirrer speed and pH by the addition of acid and alkali. Signals from the sensors are amplified and fed directly into the computer. The software decodes the signals and takes corrective action if the parameters have deviated from their setpoints. Responses are triggered by the software via digital or analogue output lines, as appropriate.

achieved using older instrumentation are often present in the software, for example on-screen help facilities, data logging and analysis routines.

DDC control does of course have disadvantages. Computers and disc drives are sensitive to mains noise and are extremely vulnerable to mains failure. Mains noise can cause data corruption or, if severe, total system failure but this can be prevented with suitable mains filters. Uninterruptable power supplies can guard against mains failure, however they are expensive and their necessity is a matter for individual judgement. The hardware configuration requires that the computer system and interfaces are placed close to the fermenters themselves in order to prevent signal degeneration over long cabling distances. There is a consequent risk of damage to the system since it may be present in an environment that is unsuitable for electronic equipment. A limited number of fermenters can be controlled by a single microcomputer since DDC software puts a very high demand on the microprocessor which would be excessively slowed by too many control loops.

5.2 Supervisory setpoint control

A typical supervisory setpoint control (SSC) system consists of one or more fermenters each with a number of intelligent independent controllers, termed process, multiloop or front-end controllers. The controllers are connected via a data cable to a computer. A typical SSC system is illustrated in *Figure 3*. Each controller can handle a number of control loops, eight is a typical number. Modern controllers can be operated directly by means of push buttons on their front panels and controllers normally possess their own data readouts. Thus each fermenter is capable of operating independently of the computer, whose role is confined to data collection at regular intervals and to instructing the controllers to change setpoints as required. The computer is therefore said to be performing a supervisory role. There are many controllers available and the purchase of suitable controllers is something that demands careful consideration. For example, some controllers need their own individual private cable to the central computer while others can use a bus system, whereby the controllers are connected to the same cable but only respond to signals addressed to themselves. The complexity of the commands to configure and install controllers can vary significantly, as can their costs, speed of operation and accuracy. Furthermore, it is always necessary to ensure that the controllers will communicate with your chosen computer and the fermenter control software!

SSC systems have a number of advantages over other methods of fermenter control. Many fermenters can be supervised by one computer because the controllers are making the second-by-second adjustments and data logging. The computer is therefore spared the time-consuming task of control and can communicate more widely since it only needs to talk at intervals to any individual controller in order to collect the latest batch of data or to change a parameter when instructed to do so by the user. A major advantage of SSC systems is their insensitivity to component failure. If a single fermenter controller fails the others are not affected. More significantly, if the supervisory computer fails the fermenters carry on regardless and can be monitored and regulated from the controllers themselves. At the very worst some data may be lost. The sensitive computer and peripherals can be separated from the hostile environment of the fermenter room since the accurate transmission of a large number of signals to and from the computer is not required. A further advantage is that the control of an entire fermenter suite can be performed from a single computer.

The disadvantages of SSC include the relatively high cost of controllers and their associated cabling. Furthermore, if the computer is located a long way from the fermenters themselves, control facilities and display screens may need to be duplicated for local viewing. It is always essential to ensure that the computer system chosen can handle the very large amount of data it may acquire from a suite of fermenters and this may require a large amount of disc storage space.

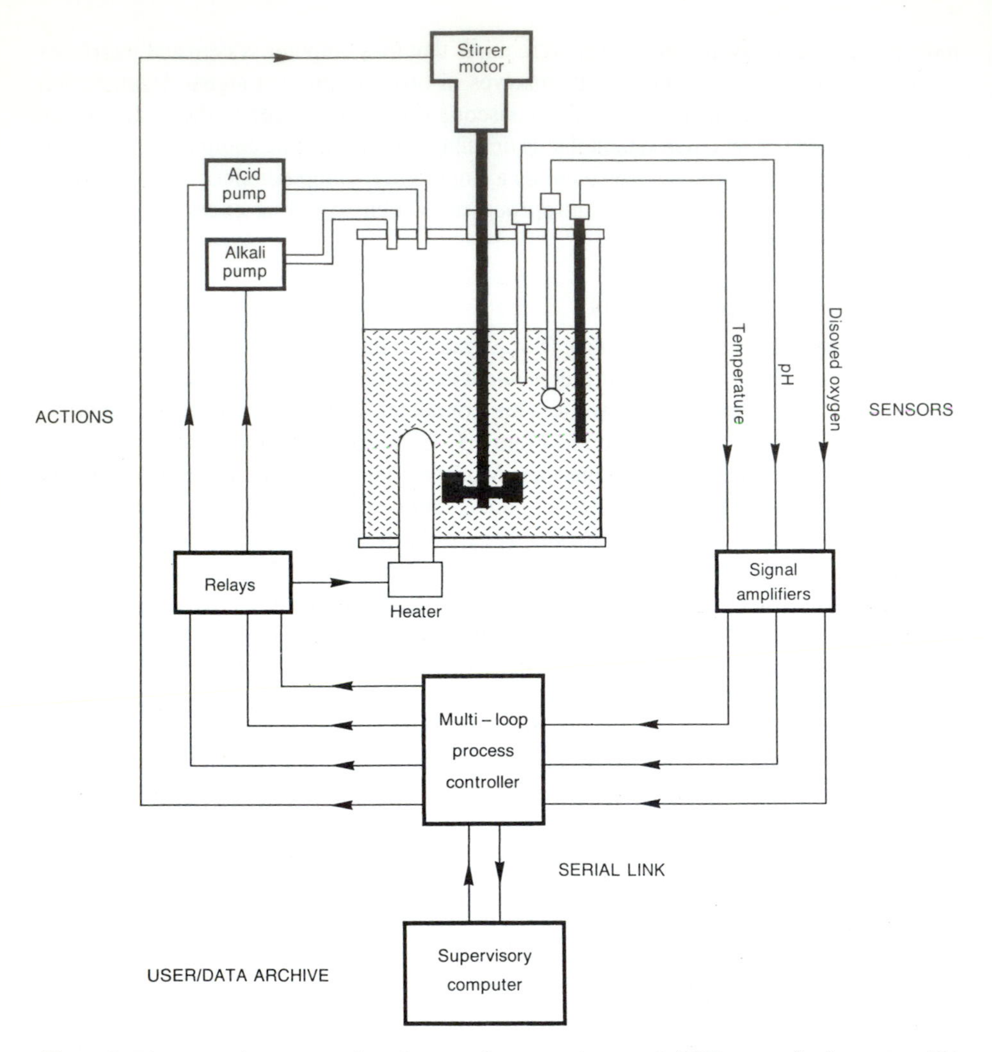

Figure 3. Diagrammatic representation of a supervisory setpoint control (SSC) system for fermenters. This example illustrates a system controlling temperature by means of heating only, dissolved oxygen tension by stirrer speed and pH by the addition of acid and alkali. All control functions are performed by the intelligent process controller and the computer only communicates with this in order to log data and send new setpoints when instructed to do so by the user.

5.3 Applicability of control systems

The type of computer control system required will depend on a number of factors, most commonly budget, size of fermentation facility, location, utilization of existing instrumentation and any specific user applications which may be required. A DDC system is more likely to be used on a single fermenter. Typically DDC systems are provided with something like eight input and 16 output channels which may seem excessive at first. However, it is always essential to allow for future expansion, otherwise short-term economy may result in the system becoming redundant as additional facilitites are requested by users can only be met by the purchase of new equipment. If capacity

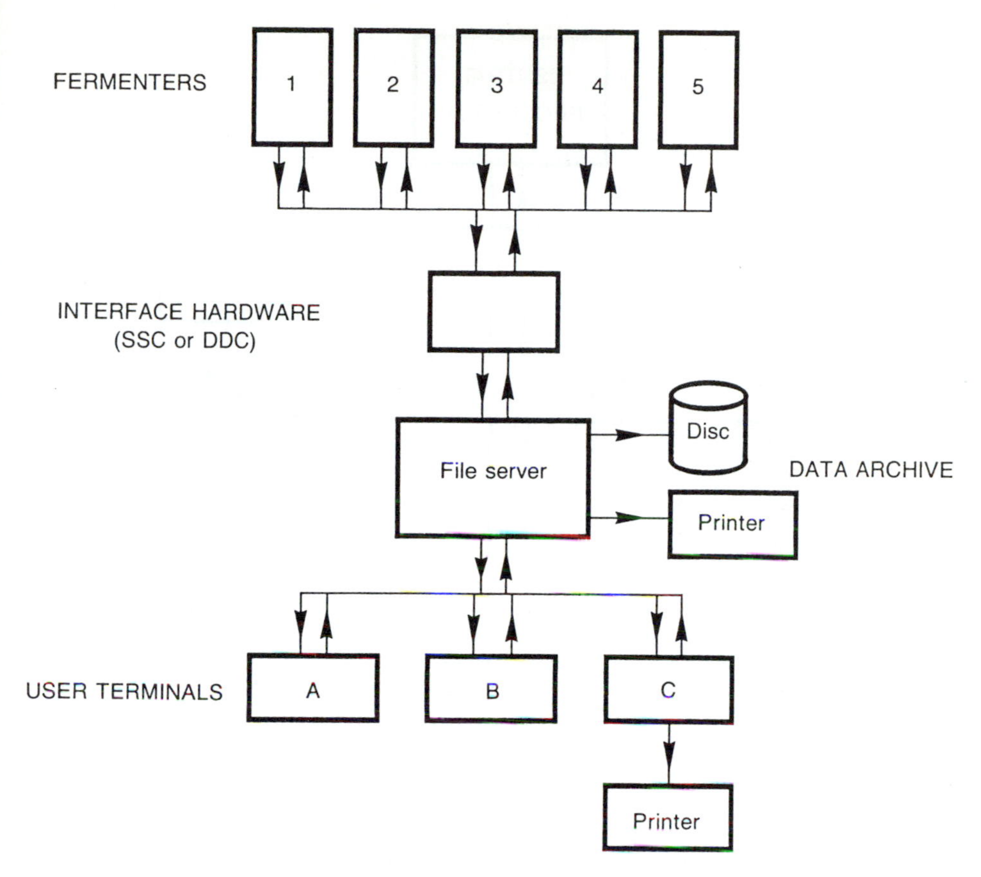

Figure 4. Diagrammatic representation of the structure of a microcomputer network system that could be used for the control of fermenters. The control of the fermenters can be achieved using either DDC or SSC configurations linked to a central computer unit termed the file server. This functions as a central storage and, in certain systems, microprocessor facility for a number of remote terminals. Since these can interrogate the file server independently the use of networks provides a means for multi-user access to a central fermentation facility.

still seems excessive, economy can be achieved by utilizing a software package that makes provision for a second fermenter to be linked to spare interface channels. This approach may, however, limit flexibility by placing constraints on the separate use of the two fermenters. If there are more than two fermenters and new instrumentation is required then SSC systems become competitive. These are advantageous because they economize on computer costs and the system can be run independently of the computer itself. However, in recent years the cost of microcomputers and mass storage devices has fallen to a level where a computer on every fermenter is a reasonable proposition. In addition, the use of dedicated DDC systems on every computer is preferred by some users in view of their increased flexibility. However, this approach can make no provision for centralized facilities such as data storage.

Both types of control system can be employed in computer networks. In a network a number of microcomputers (normally up to 255) are linked together to a powerful central processing and storage unit, the file server (2,6,7). Thus relatively cheap

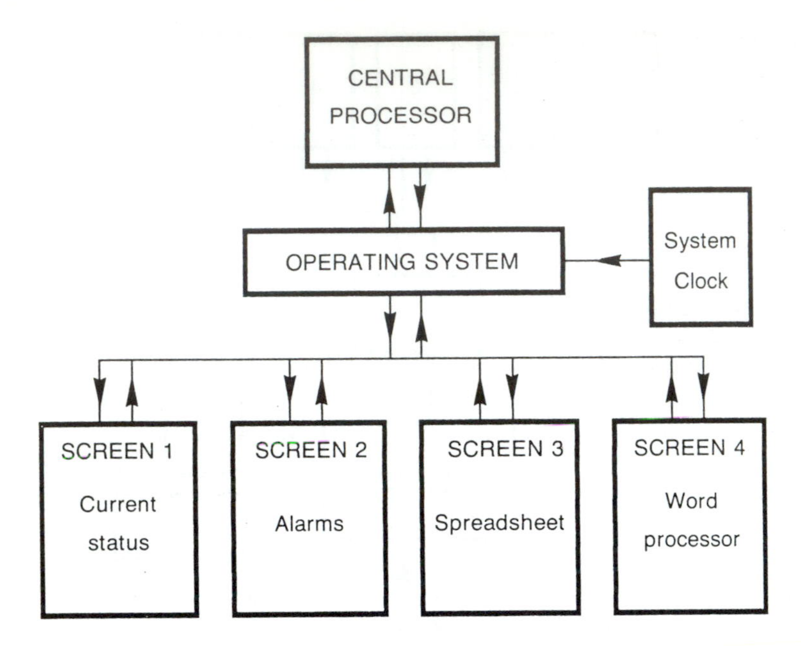

Figure 5. Diagrammatic representation of the operation of a multitasking operating system. The operating system splits program access to the computer processor into time slices. Although the user can only communicate with one program at a time (on the 'screen' or 'window'), all the programs are active and, since the cycling time between programs is so rapid, it appears to the user that the individual programs are running simultaneously. In the case of fermenter control a user could be analysing archived data or word processing whilst the computer simultaneously and independently controls fermenter parameters and logged data.

individual computers can have common access to relatively expensive central facilities. Example low-cost commercially available products for microcomputers include ECONET for Acorn/BBC computers and PC-Net for IBM PCs and compatibles. In a fermentation system, the file server could control the fermenters and log data and could be interrogated remotely by individual users (*Figure 4*). When not being used for accessing the fermenter control software, the individual terminals can be used for any purpose. This is generally a cost-effective way of spreading computer power if there is a large number of potentially conflicting users. If DDC systems are preferred, the use of a network provides a means of operating a central fermenter control and data storage facility. However, it should be remembered that dedicated fileservers and network adapters will add significantly to the total computer system cost.

The central computer in an SSC system spends much of its time doing nothing and processing power is therefore wasted. The advent of multitasking operating systems for personal computers provides a way of overcoming this problem (8). The principle of multitasking operating systems is illustrated in *Figure 5*. The operating system can split the microprocessor's time between several programs at once. This process is so rapid that the programs appear to the user to be operating independently and simultaneously. Consequently, the computer can be utilized for other tasks, such as word processing or post-run data analysis, whilst supervision of the fermentation controllers continues.

To summarize, DDC systems are most commonly employed in small research environments where their flexibility and ease of use is more important than central monitoring and absolute reliability. SSC systems have the advantage were large numbers of fermenters are operative, where a centralized facility is preferred or where reliability is essential. The ultimate choice is personal.

6. SOFTWARE

We have discussed all of the hardware and techniques that may be involved in computer control of a laboratory fermentation suite. Once these factors have been considered by the user and he or she has decided on the type of system to be installed it is necessary to consider the computer software itself. It is always advisable to get a number of software packages demonstrated and to test the features for oneself. There are two basic types of software that can be chosen, namely dedicated fermenter control software and general process control software. The former is written with the microbiologist specifically in mind and packages vary considerably in quality, price, complexity and flexibility. This type of package is available for both DDC and SSC control systems. Process control software is written for general control operations, most commonly for industrial applications. The software is suitable for SSC control only and will need to be specifically tailored, normally by the user, to fermentation systems. This may be a time-consuming and difficult task. This type of software does have the advantage, however, of being extremely flexible and enables the user to obtain a system tailored to his or her specific needs.

The type of software required depends very greatly on the fermentation system, the people who will be operating it and the system manager's experience and desire to make alterations. Remember that in the end it is the software and not the control hardware that makes the system usable and it is sensible to examine a range of software packages from different suppliers before making a final choice. Besides ensuring that the software is capable of controlling your chosen system and has sufficient capacity for future expansion there are a number of features which it may be useful for the software to possess. As a guide for the potential purchaser this section discusses some of the features that need to be considered when choosing fermenter control software.

6.1 **The user–computer interface**

It cannot be emphasized strongly enough that the system will ultimately be operated by people who know little about computers and fermenter control but need to be able to operate the system quickly and easily. The user–computer interface is therefore a very important consideration in choosing software. Ease of use is something that is very difficult to define and is highly subjective. In general, people like menu-driven programs with clear instructions and well designed screen displays. Other features that are desirable include on-line help, easy exit from procedures entered in error and checking for incorrectly entered values. Colour displays are not essential but make for clearer screen layouts and generally make things easier to understand. Clear instruction manuals are extremely useful for both the novice and the expert, however, many commercial manuals are anything but well written.

Data handling is an important consideration. Facilities for adding off-line data (e.g.

cell counts, ATP concentration, substrate analyses) and performing on-line calculations (e.g. oxygen uptake rate, mass balances, specific growth rate) may be available as part of the fermenter control software or can be performed as part of post-run data analysis. Details of the commonly employed derived calculations have been documented previously (9).

Most commercial packages will have some post-run analysis built in, for example simple spreadsheets are commonly provided. For dealing with large amounts of data it is normally faster and more convenient to convert the data files to a standard spreadsheet format and use the far more powerful commercial packages, for example Lotus 1-2-3 on the IBM PC or Viewsheet on the BBC microcomputer. Once the data is in a spreadsheet, selected portions can be printed, off-line data added, calculations performed and the data can be reduced to smaller sets by averaging groups. Merging of data from different runs is also possible. The ability to select data points is important if the data is to be printed out using commercial graph plotting packages which can only handle 50 – 100 points at a time. To make the plotting process easier an integrated package with spreadsheet and graph-plotting facilities combined is highly recommended.

6.2 **Some useful features**

There are a number of additional useful facilities that may be present in software and may be of particular benefit to some users.

6.2.1 *Security*

In certain environments security facilities may be useful in order to prevent accidental or deliberate alteration of parameters without the need for physically locking the machine away. This is normally achieved by the use of passwords to the system itself or to certain subsections of the software.

6.2.2 *Time stamping*

Some systems provide a time stamping function whereby any modifications to parameters or changes in status are recorded on disc and printer with the date and time at which they occurred. A similar feature is the ability to enter multiline comments during a run, which are time and date stamped and stored on disc or dumped to the printer.

6.2.3 *Alarms*

The provision of alarms is a common feature. If user-defined limits are exceeded an alarm message is issued to screen, disc log and/or printer. Facilities are often provided to trigger audible or visual alarms to that action is taken if the alteration in the parameter requires urgent remedial attention.

6.2.4 *Time scheduling*

This type of facility enables the user to set up a series of operations to be performed at certain times. For example, nutrients could be added at certain times or data can be logged to disc in timed blocks if experimental purposes require this.

6.2.5 *Sterilization routines*

Some fermenters can be sterilized *in situ* by means of high capacity heaters or a steam supply. These processes are readily controlled by software.

7. FERMENTER CONTROL SYSTEMS—CASE HISTORIES

Since there are so many possible combinations of fermenter systems, computers, interface hardware, control software and user requirements, it has only been possible to discuss in general terms the factors that may require consideration in choosing and installing a fermenter control system. In order to demonstrate some practical aspects of the process, this section describes experience the authors have gained during the choice, installation and operation of two control systems, one SSC and one DDC. No specific endorsement of any products is intended.

7.1 **SSC system**

This system was purchased to link to an existing group of five fermenters used for batch and continuous culture and was intended to provide more accurate control than an existing system, to enable data logging and to be sufficiently flexible and powerful to enable it to be used for the control of at least five more fermenters. In view of the requirement for the control of these from a single computer, only an SSC system could be considered and this had the added benefit of preventing excessive reliance on the computer hardware and software. Five parameters were to be controlled on each fermenter:

(i) pH, by the addition of acid and alkali;
(ii) temperature, by means of a heating element and a valve-controlled cold water 'finger';
(iii) dissolved oxygen, by gas blending;
(iv) foam, by the timed addition of anti-foam;
(v) stirrer speed.

However, it was felt that it was important to provide the capacity for future expansion of control requirements. The entire process was simplified by the fact that on-site electronics engineers experienced in fermentation control systems were involved at all stages of the selection process and installed the system hardware.

Although there were many suitable process controllers that could have been used, previous experience enabled the choice of process controllers of known quality. A programmable 8-loop controller was provided for each fermenter. This provided enough capacity for future expansion of control requirements and had the added benefit of ensuring that the failure of one controller would not interfere with more than one fermentation. The controllers have a digital display on their front panels and are capable of being programmed by means of push buttons and therefore can easily be operated independently in the advent of a computer failure. The choice of signal amplifiers for the sensors and other interface electronics was also simplified by previous experience and care was taken to provide suitable matching of individual input and output signals in order to obtain the highest accuracy of control. Having chosen the electronics it was then necessary to select suitable software. Two options were available: a general process control system or a dedicated fermenter software package. Demonstrations were arranged of a number of potentially suitable packages and the more favoured were reviewed twice. It was felt that, in this particular case, the dedicated fermenter control software packages were somewhat lacking in expansion possibilities and, particularly, flexibility. As a consequence of the demonstrations and extensive discussion both with end-users and those experienced in fermenter control systems, a process control package was purchased.

The requirement for software power and flexibility was paramount and a consideration of some of the features of the package may illustrate why this particular system was chosen. The computer hardware required for the software version purchased is an IBM PC/AT with an enhanced colour graphics card and a mathematics co-processor. The operating system used is Concurrent DOS, which is multitasking (see Section 5.3). Multitasking is necessary due to the modular nature of the software which requires several programs to run simultaneously. This has the added advantage of enabling other tasks to be performed on the computer while fermenter control continues. The main module of the control software is the user–machine interface. Display screens can be designed as necessary using a drawing package. The software uses data tables for values of digital inputs, digital outputs, analogue inputs and analogue outputs which can be accessed and printed as required on any screen display. Meaningful names can be linked to the data table locations, as can alarm values and trends of data over a user-defined period. Logging of individual values can readily be achieved and the data is stored on hard disc or floppy disc, as required. Every aspect of the system is user-definable and the software can therefore be configured for a specific use and altered or expanded at will. An essential program module that operates as a background task communicates with the controllers. This program checks output data tables for changes and sends the information to the controllers and reads current values from the controllers which it places in the data tables for access by the user program. A number of other modules are provided that operate in other windows of the operating system. A simple spreadsheet package is supplied which can display selected current and archived data and perform all normal spreadsheet calculations upon the values. The data may also be plotted, graphed or converted to other formats for use in more powerful spreadsheet packages such as Lotus 1-2-3. There is also an alarm historian that displays the last 100 alarms on screen and a time schedule option that enables the triggering of functions at a specific time. This latter program has been used for the addition of a nutrient pulse at certain stages during fermentation. For day to day use, other software is present on the computer for use in other windows, including a word processing package.

Having purchased the hardware and software, the installation process could proceed. Due to in-house expertise, installation of the interface electronics and the controllers was achieved with little difficulty. Configuration of the controllers was also relatively simple, although it was necessary to spend some time testing and adjusting the individual loop parameters in order to ensure that optimal control was achieved. This was particularly important for foam control. Initial loop settings resulted in the addition of anti-foam once foaming had been detected but insufficient time was allowed to enable the anti-foam to act before the next addition occurred. This resulted in the addition of excessively large quantities of anti-foam which inhibited culture growth and, due to the high viscosity of the solution used, prevented efficient oxygen distribution through the fermenter. Careful adjustment of the loop enabled a suitable delay period to be selected which prevented overdosing with anti-foam yet did not permit excessive persistence of foam if the initial anti-foam addition was not effective. Having completed the control aspects the next step was the installation of the computer system. Communication between the computer and the controllers is by means of a serial link and this was relatively easily established. However, software installation provided some problems. It was found that the program that communicates between the user screens and the controllers was not capable of placing the decimal points at the correct position within the logged value, even though the controllers were behaving correctly. The problem was referred to the suppliers and thereafter to the parent company in the USA where the software was written. An error in the program was discovered and the problem cured, but several weeks were wasted. Having established communication between the software and the controllers, it was necessary to design the user screens, to link the data to the displays and to program the alarms and data trends. This process was relatively simple once mastered but was extremely time-consuming and hindered by a manual that was virtually incomprehensible. No further major problems were encountered, although some re-programming of a portion of Pascal source code was necessary to get the software to drive the chosen printer.

At the time of writing, the entire system has been operating effectively for several months and experience has highlighted a number of factors that should be taken into account when choosing an SSC system. Extensive consultation is essential and it is wise to see a number of rival systems demonstrated before making a choice. The end-user who, it must be remembered, is unlikely to be a computer expert likes clear screens, menu-driven operations, on-screen help and a simple instruction manual. The latter is often best written in-house by the system manager. The experiences outlined above illustrate some of the problems that are likely to be encountered in the installation of a control system even when carefully chosen. The complexity of the software in this case history is a disadvantage resulting from the requirement of total flexibility, since control of all aspects of the software will, by definition, be complex. The use of a specific fermenter control package would undoubtedly have minimized the complexity but may have hindered future development. Nevertheless, the programming error caused severe problems and should not have occurred. However, this is unlikely to happen frequently! The hardware aspects of the system have functioned excellently, thanks to the efforts of the electronics engineers, and this emphasizes the importance of extensive discussion during the decision-making process and care during installation. The requirements and installation of the fermenter control system described are certainly more complex than many others and this will therefore have presented greater difficulties than many people will encounter. However, this case illustrates that, with advice, assistance and careful deliberation, an SSC package may represent a good choice for the control and operation of a number of fermenters.

7.2 **DDC system**

The system described below was originally conceived to provide control of two LH2000 series fermenters, one fitted with a 5-litre vessel and the other with a 15-litre vessel, and was required due to the absence of any suitable commercial packages. The software and interface were produced in the Department of Biological Sciences, University of Warwick during 1986.

The computer system originally chosen was an Acorn BBC model B microcomputer since it was inexpensive, powerful and in widespread use. A 20 K RAM expansion board was fitted and the computer was linked to two floppy disc drives, a colour monitor and an Epson dot matrix printer. The original computer has subsequently been replaced by an Acorn Master 128 machine. Both fermenters were to be controlled for the following parameters:

(i) pH, by means of pumped acid and alkali addition;
(ii) temperature, using 2 kW heating elements and valve-controlled cold water fingers;
(iii) dissolved oxygen by means of stirrer speed control or the regulation of air flow through the vessel.

Both fermenters were capable of being sterilized *in situ*, the smaller vessel by means of the heating element and the larger one by means of a steam supply. The interface board for the system was purpose-built and was designed to fit into the standard LH rack casing. The interface provides eight 10-bit analogue inputs, eight digital outputs and one 8-bit analogue output. Connection to the computer is via the 1 MHz bus port (a high speed parallel interface) on the microcomputer. The standard LH2000 amplifiers are used to condition the signals from the probes and output from these is in the form of a 0–5 V signal that can be directly linked to the analogue input socket fitted on the microcomputer. Control of pH and temperature is achieved using the digital outputs to switch pumps, heaters or water valves. The digital signals drive 24-V relays that in turn feed 240-V mains supply to the relevant pieces of hardware. The control of dissolved oxygen is achieved using the analogue output port. The stirrer speed can be varied directly by the 0–5 V signal produced. However, some organisms cannot stand the high shear forces induced by rapid stirring and control of oxygen in these cases must be attained by controlling gas flow. To achieve this, the analogue output port is wired to a 0–5 V stepper motor that drives a needle valve. In this case, the stirrer motor is not connected to the computer and must be operated directly from the front panel of the fermenter.

The control software was designed to be highly user-configurable, so that anybody could fine-tune the operation of the fermenter. The user can label the input and output channels with meaningful names and can link any input channel with any two digital channels or the analogue output channel in order to achieve control. Fine adjustment of the proportional control is achieved by means of an interactive graphical display of the gain profiles of the output channels. Once the control loops have been programmed, the user can calibrate the signals from the probes, enter setpoints and set alarm values. The setup data can be saved to floppy disc as a named parameter file and re-loaded whenever required.

A separate program is used to control the sterilization routines. Direct control of the heating element or a valve on a steam supply is achieved by means of a digital output signal. One problem that occurred was that the temperature probes provided an 8-V signal at 121°C but operated in the 0–5-V range when measuring fermentation temperatures. This 8-V signal needed to be attenuated before it could be matched to the 0−5 V range of the ADC but this would have significantly reduced the accuracy of the signal measured at operating temperatures. In order to avoid this problem, a separate, attenuated channel was used to log sterilization temperatures, a clear illustration of the usefulness of specifying more interface channels than may at first appear necessary.

In operation, the system had proved reliable and acceptable to the users. Only one major problem was encountered: the frequent crashing of the computer due to mains interference. This was significantly minimized by the purchase of a mains conditioning unit to provide a clean power supply, but even so a few crashes still occurred. These could be extremely problematical since they sometimes occurred during a control response. For example, a crash that occurred while the acid pump was running resulted in the pump remaining continuously on. To prevent this happening a simple timer was added to the interface which is re-set periodically by a signal from the computer. If no signal is received within a 10-sec period, the timer alerts the interface and all digital outputs are shut down until a new signal is received from the computer. The analogue output is not regulated in this way in order to ensure that oxygen supply is continuously maintained. The original model B microcomputer was also provided with battery-backed RAM that enabled the software to automatically re-start after a total power failure. Battery-backed RAM is fitted as standard to the Master 128 machine and serves to perform the same function.

8. CONCLUSIONS

We have discussed in some detail all of the aspects of computer control of fermenters that need to be considered when purchasing a system. Obviously, requirements vary greatly from case to case and it is not possible to make specific recommendations as to what is best. When considering fermentation control, whether brand new or as a bolt-on upgrade to existing apparatus, it is necessary to think very carefully about what is required, how accurate control needs to be and, most importantly of all, how the routine user will react to the system. It is certainly of great benefit to be able to approach commercial suppliers with some knowledge of what you want and why.

It is possible to make a few generalizations on how to proceed. Always consult widely about what is needed since money spent on unnecessary accuracy or facilities is wasted and could have been better spent on another aspect of the system. However, always attempt to build some capacity for later expansion into the computer hardware, software and the interfaces. Once users come to realize the capabilities of a well thought out computer control facility they are likely to request additional features. If adequate computing power, storage facilities, software capability and interface loops are provided these can be implemented at minimum additional cost and effort. Finally, be very careful about choosing software. A well thought out package should be easy to use and will therefore save both the system manager and the user time and effort. State of the art control equipment linked to substandard software is useless since it will never be used.

9. ACKNOWLEDGEMENTS

We would like to thank Dr Andrew Turner, University of Warwick and Dr Ruth Morgan, Wellcome Biotech, Beckenham for their assistance in the production of this chapter.

9. REFERENCES

1. Armiger,W.B. and Humphrey,A.E. (1979) In *Microbial Technology.* 2nd edn., Peppler,H.J. (ed.), Academic Press, London, Vol. 2, p. 375.
2. Rolf,M.J. and Lim,H.C. (1986) In *Comprehensive Biotechnology. Volume 2. The Principles of Biotechnology: Engineering Considerations.* Humphrey,A.E. and Conney,C.L. (eds), Pergamon Press, Oxford, p. 165.
3. Bull,D.N. (1986) In *Comprehensive Biotechnology. Volume 2. The Principles of Biotechnology: Engineering Considerations.* Humphrey,A.E. and Cooney,C.L. (eds), Pergamon Press, Oxford, p. 149.
4. Titus,J.A., Luli,G.W., Dekleva,M.L. and Strohl,W.R. (1984) *Appl. Environ. Microbiol.*, **47**, 239.
5. Wimpenny,J.W.T. (1985) *Microbiol. Sci.*, **2**, 53.
6. Morgan,P., Whiteside,M.C. and Dow,C.S. (1985) *Binary,* **4**, 13.
7. Whiteside,M.C. and Dow,C.S. (1984) *Binary,* **1**, 16.
8. Morgan,P. (1987) *Binary,* **10**, 20.
9. Zabilske,D.W. (1986) In *Comprehensive Biotechnology. Volume 2. The Principles of Biotechnology: Engineering Considerations.* Humphrey,A.E. and Cooney,C.L. (eds), Pergamon Press, Oxford, p. 175.

10. APPENDIX

Fermenter control software

We have listed below some suppliers of the various kinds of software that can be employed for fermenter control. The list has been restricted to software that can be used with a range of fermentation systems and we have deliberately not included products from manufacturers that can only be used with their own hardware. It has not been possible to give full details of the products and many suppliers can offer a range of packages—they should be contacted directly if you are in the market for their type of system. The list is far from exhaustive—we apologise to those suppliers we have overlooked and to those whose products may not have been described effectively.

LH Fermentation, Bells Hill, Stoke Poges, Slough SL2 4EG, UK. A range of systems for a variety of microcomputers including the BBC micro.

Alfa-Laval, Great West Road, Brentford, Middlesex TW8 9BT, UK; 2115 Linwood Avenue, PO Box 1316, Fort Lee, NJ 07024, USA. A range of systems for many makes of computer and both single and multi-fermenter uses. Market a powerful BBC micro-based DDC package.

Rintekno oy, PO Box 146, SF-02101 Espoo, Finland. Multi-fermenter control software for DEC PDP11 minicomputer. Suitable for research, process development and production fermentation.

New Brunswick Scientific, 6 Colonial Way, Watford WD2 4PT, UK; Box 4005, 44 Talmadge Road, Edison, NJ 08818, USA. Intelligent process controller specifically designed for fermenter control that can handle multiple fermenter rigs and can be interfaced to many computers. Also market an IBM PC-based package for SSC control.

Biochem Technology, 66 Great Valley Parkway, Malverne, PA 19355, USA. IBM PC-based package specifically designed for DDC operation of fermenters. Will handle up to 20 on-line variables.

Camm Technology, 7 Hazelwood Close, Dominion Way, Worthing, West Sussex BN14 8NP, UK. SSC process control systems for IBM PC. Can be employed for fermentation.

Drew Scientific, 12 Barley Mow Passage, Chiswick, London W4 4PH, UK. DDC system employing dedicated computer specifically designed for fermenter control. Multifermenter capability.

Amber Systems, The Venture Centre, The Science Park, Coventry CV4 7EZ, UK. DDC and SSS systems employing a dedicated computer. Multitasking operating system. Multifermenter capability.

Bioengineering AG, Sagenrainstrasse 7, CH 8636 Wald, Switzerland. DDC system employing a dedicated computer. Multifermenter capability.

Biotechnology Computer Systems, Cleveland House, Church Path, Acton Green, Chiswick, London W4 5HR, UK; 3563 Arden Road, Hayward, CA 94545, USA. DDC IBM PC-based system for single fermenter use. Multifermenter control system for DEC Microvax and PDP11 computers.

Dian Microsystems, Mersey House, Battersea Road, Heaton Mersey, Stockport, Cheshire SK4 3EA, UK. Process control software that can be configured for SSC multifermenter control. Runs on DEC PDP11 minicomputers and IBM PCs. Multitasking operating system.

HCS Industrial Automation, Populieranlaan 3, Postbus 88, 8050 AB, Hattem, The Netherlands. Process control software that can be configured for SSC multifermenter control.

Hewlett-Packard, King Street Lane, Winnersh, Wokingham RG11 5AR, UK; Mailstop 20B3, PO Box 10310, Palo Alto, CA 94303, USA. Process control system that can be configured for SSC control of fermenters. Runs on HP Vectra, HP1000 and IBM PCs(?).

Turnbull Control Systems, Broadwater Trading Estate, Worthing, West Sussex BN14 8NW, UK; 11515 Sunset Hills Road, Reston, VA 22090, USA. Process control package for IBM PCs that can control fermenters by SSC.

INDEX

Page numbers in bold denote main entry